American Romanian Academy of Arts and Sciences (ARA)

California, USA

METALLIC ALLOYS FOR ENGINEERED APPLICATIONS

Ildiko PETER

ARA Publisher International Academic Press

Title: Metallic Alloys for Engineered Applications
Author: Ildiko Peter
Editors: Ruxandra Vidu
Cover: Illustrations provided by Ildiko Peter

Published by
ARA Publisher Academic Press, an International Publishing House of the American Romanian Academy of Arts and Sciences,
University of California Davis,
http://www.AmericanRomanianAcademy.org
Address: P.O. Box 2761
Citrus Heights, CA 95611-2761

Copyright © 2018 by American Romanian Academy of Arts and Sciences

ISBN 978-1-935924-28-9

No part of this publication may be reproduced, stored in a retrieval system or transmitted in any form or by any means, electronic, mechanical, photocopying, recording, scanning or otherwise, without the express prior written permission of the Publisher.

LIMIT OF LIABILITY/DISCLAIMER OR WARRANTY: The publisher make no representations or warrantees with respect to the accuracy or completeness of the content of this work and specifically disclaim all warranties, including without limitation warranties of fitness for a particular purpose. No warranty may be created or extended by sales or promotional materials. The advice and strategies contained herein may not be suitable for every situation. This work is sold with the understanding that the publisher is not engaged in rendering legal, accounting, or other professional services. If professional assistance is required, the services of a competent professional person should be sought. Neither the publisher nor the author shall be liable for damages arising therefrom. The fact that an organization or website is referred to in this work as a citation and/or a potential source of further information does not mean that the author or the publisher endorses the information the organization or website may provide or recommendations it may make. Further, readers should be aware that internet websites listed in this work may have changed or disappeared between when this work was written and when it is read.

ARA Publisher also publishes its books in a variety of electronic formats and by print-on-demand. Not all content that is available in standard print version of this book may appear or be packaged in all book formats.

Table of Contents

Outline 5

1. Introduction 6

2. Classification of materials 8
    2.1. The crystalline structure of metals 8
    2.2. Extraction and processing of minerals 10

3. Aluminum and aluminum alloys 13
    3.1. Foundry processes 18
    3.2. Considerations on the presence of defects in casting 19
    3.3. Heat treatments for Al alloys 25
    3.4. T6 and T7 heat treatment of Al-Si-Cu-Mg alloys 34

4. Die-casting 35
    4.1. Some consideration on the use of dies for manufacturing components 39
    4.2. Thermal treatments of steels for dies 44
    4.3. Thermoregulation and mold lubrication 48
    4.4. Some considerations on the defects on castings 54
    4.5. The technological degradation of the dies 59
        4.5.1. Influence of process parameters 64
        4.5.2. Thermal Fatigue 66
        4.5.3. Thermal stress on the surface 68

5. Hot forging and some recommendations during its application ... 69
    5.1. Plastic deformation and forging 70
    5.2. Stages of the hot forging process 74
        5.2.1.   Cutting the billet 75

    5.2.2. Heating of the material 76
    5.2.3. Molding operation 77
    5.2.4. Subsequent processing (deburring, tooling) 78
    5.2.5. Thermal treatments (if any) 78
    5.2.6. Blasting 78

6. Some considerations on X-Ray diffraction and its use for metals and alloys - Examples of real case studies 82
    6.1. Generalities 82
    6.2. Real case studies 87

7. Concluding remarks 97

References 98

Outline

The aim of the present work is twofold. On one hand in the first part, the attention is focalized to give a brief but general idea on (i) the presence and importance of the different materials which surround us, including light alloys and more in details Al-based alloys, on (ii) some features on the actual state of the manufacturing of metallic alloys, and on (iii) some technical hitches appearing when metallic alloys are developed. On the other hand, some indications on how the metallic materials can be structurally characterized using complementary methodologies, namely microstructural analysis and X-Ray diffraction technique, including different real case studies, will be recalled. Additionally, a few aspects on how the data acquired can be managed in order to recognize the highest number of information bits about the investigated material and to holistically use them for finding a suitable application, or vice versa to find the right material for application drive processes, will be discussed.

The present book is mainly addressed to students of materials and mechanical sciences, electrical and industrial engineering, and other relevant engineering specializations, where investigations on materials/metallic alloys are important.

1. Introduction

Materials make up the substances, which surround us. Commonly, one can talk about traditional materials such as wood, concrete, bricks, steel, plastic, aluminum etc., or new materials, which are growth as a result of constant research and development.

On the basis of their composition, the materials can be classified as polymeric materials, ceramic materials and metallic materials. In order to be able to make a selection between various materials for a specific industrial application, one has to know the "life" of the materials, including at least their production and post-production route types as well as the macro and microstructural, thermal and mechanical properties.

The selection to discuss about Al-based alloy is for the reason that actually such materials are very challenging, since they represent the best choice to produce cast parts in the transport industry and to replace Fe and steel components in some other even strategic areas such as aerospace and others. In such application fields there is a clear tendency to employ lightweight alloys, because it allows decreasing the total weight of the vehicles that in turn leads reducing fuel consumptions and emissions. Fuel consumption and exhaust emissions are directly related to the car weight and for that, the automotive industry is looking for innovative process technologies which can allows the production of components of light alloys with improved properties.

Other important area of application related to Al-based alloys is the production of metal matrix composites, because of their low specific density, better oxidation resistance in comparison to Mg and excellent ductility if compared to Be and other light metals.

Reinforcing ceramics, in the form of fibers, whiskers or particulates of oxides, carbides and nitrides, leads to a variety of properties and offers an excellent combination with the matrix properties.

Other issues which have been considered in this monograph are concerning the most important difficulties which one meets during the manufacturing process. The presence of defects, the choose of the right tools for the production, the direct control of the different parameters during the production route, are only some of the aspects that one has to be take into account when the development of high quality product is the main goal. Some reflection on the defects which are generated during the production, the use of dies and the related issues which are created during their use and how one can solve the problems appeared, or how one can act or overpass their appearance. The following task direction is related to hot forging which actually has traditionally represented one of the most important processes in the metal working sequence, because forged components have always offered maximum reliability and superior mechanical properties. However, the gap between the performance of components forged and obtained through other processes has been gradually reduced as a result of the improvement of other techniques. Furthermore, the increasingly pressing economic conditions, pollution controls and world competition have had a major impact on the forging process.

Moreover, useful, representative information on some real case investigations will be presented on how the structure of metallic materials can be described using X-Ray diffraction technique.

2. Classification of materials

Science of Metals can acquire skills on the basic principles that determine the various chemical, physical and metallurgical phenomena that are involved in the design stage and in the actual management of metallic materials.

Metallic materials are inorganic materials made up by one or more metallic elements (Fe, Cu, Al, Ni, Ti, etc.), which can also contain some non-metallic elements (C, N, O, etc.).

Metallic materials have a crystalline structure where the atoms are arranged in a regular way in space. Generally, metals are good thermal and electrical conductors, they are relatively resistant to mechanical stresses, shows ductile properties at room temperature. Many of them maintain an adequate strength even at high temperatures.

1. Metals and ferrous alloys: they contain a large percentage of Fe: e.g., steels and cast irons;

2. Metals and non-ferrous alloys (Al, Ni, Zn, Cu, Ti): they do not contain Fe, or contain only an insignificant quantity of such elements.

2.1. The crystalline structure of metals

Metals are characterized by the presence of the typical metallic bond. It has a delocalized character and it consists of an electrostatic attraction between the positive metal ions and the mobile conduction electrons within the whole structure.

Metals and their alloys are generally crystalline materials; however, there is a specific class of metal based materials which can be developed through a process which involves rapid cooling with a final product in an amorphous system. Metals and alloys are characterized by:

- electric conductibility;
- photoelectric effect;
- thermionic effect;
- thermal conductivity, etc.

The electrical conductivity is due to the presence of free electrons which are able to move under the action of an external electric field and can be considered as a measure of the material's ability to conduct the current. The conductivity in metals and alloys varies as a function of temperature: as temperature increases, the electrical conductivity decreases because the movement of the electrons is inhibited by the progressive intensification of the atom's oscillations.

The photoelectric effect is the characteristic that makes easier the extraction of the electrons when irradiated by an external source. It is well known that certain metals and alloys have the property of losing electrons when exposed to electromagnetic radiation of suitable energy and frequency.

This phenomenon finds widespread application in technology of everyday life (e.g. in the mechanism regulating the closing of the elevator's sliding doors).

The thermionic effect is the characteristic that makes easier the extraction of the electrons; it is the physical phenomenon which consists in the liberation of the conduction electrons by a metal to which heat has been externally supplied. Normally, the conduction electrons of a metal are free to move within the solid matrix, but do not have enough energy to fall out. When one externally provides energy under the form heat, thermionic effect takes place leading to remove electrons from the metal.

The thermal conductivity is related to the transport of the thermal energy due to the mobility of the electrons and is defined as the transport of energy within the structure as a result of a temperature gradient.

When the lattice planes can slip relatively easily over each other without any alteration of the binding interactions, one can talk about ductility and malleability. Metals are characterized by non-transparency and gloss; they absorb all incident visible radiation and re-emit in all directions. The electrons can be excited to higher energy levels by absorbing any amount of energy.

2.2. Extraction and processing of minerals

The minerals containing the metal is collected from the mines, then subject to enrichment processes and of chemical transformations leading to the oxides state which are subsequently reduced to the metallic state, using different processes depending on the nature of the oxides.

From such a reduction, impure metal alloys are obtained, which are subsequently processed and refined to achieve the technically pure metal and/ or alloys. This material can be considered as starting one for the production, at industrial level, firstly of the semi-finished products and then of the final products.

From the furnace the metal is extracted as a liquid metal and then it can be cast:

- in appropriate molds acquiring the final shape; or
- more commonly casting is carried out obtaining the ingots.

These products are usually submitted to additional specific processing techniques to obtain a semi-finished product and finally the finished metal products. Metallic alloys when submitted to thermal or thermochemical treatments an enhancement of some properties can be obtained and the development of a more resistant material takes place (e.g. increasing of the corrosion resistance, of the mechanical strength, etc.).

Mechanics of materials is a discipline which studies the mechanical behaviour of materials: leading to a good forming and to their coherent use in structural field promoting as well the development of cutting-edge tools and components.

The mechanical strength is a property indicating both the maximum stress that a common material is capable to support before fracture, and the minimum stress necessary to start its permanent deformation.

The mechanical strength of the materials to various types of stress can be measured by specific tests, e.g. traction, compression, bending, shear/torsion, impact, fatigue and creep tests. The resistance of the materials depends on various factors such as:

- the type of materials, including their composition and their quality;
- the type of the manufacturing, treatment and storage;
- the state of tension applied (uniaxial, biaxial, triaxial);
- velocity of the stress application, its duration and eventually the replication of the load;
- climatic conditions.

Under load, development of micro-holes occur which has a local character. If the material plastic deformation at local level takes place with difficulty, development of a fragile crack happens leading to a brittle fracture, while the material easier deforms plastically formation of a ductile crack takes place with a consequent ductile fracture.

When a metallic sample is subjected to a tensile test with an increasing stress, the sample firstly will linearly deform according to the Hooke's elasticity law (elastically deformation) and removing the applied stress the metal will return to its original shape. Increasing the applied stress beyond a certain limit determines an irreversible plastic deformation, a strain hardening, necking regions and finally failure takes place. If the theoretical value of the energy necessary to plastically deform the sample is significantly higher than necessary, it is really due to the presence of defects in the crystalline structure. The most important defects in this sense are a line defects or dislocations which are not only moves within the crystalline structure, but they are growing with a suitable mechanism responsible of the hardening phenomenon.

3. Aluminum and aluminum alloys

Aluminum is widely present on the Earth's crust in the form of mineral and it is one of the main constituents of all rocks: it is present in the form of silicates such as feldspars, micas and clays. Among these minerals, the most important as concerns the aluminum is the bauxite.

The high affinity of aluminum versus oxygen avoids the reduction to the metallic state with classical metallurgical methods. The use of aluminum on a large scale has become possible only since the late nineteenth century, a period in which the development of the two steps method arises, allowing to obtain aluminum metal directly from the mineral. The first one is the so called Bayer process that is an alkaline method for the extraction of alumina ($Al2O3$) from bauxite. The second one consists in the subsequent electrochemical reduction of alumina to aluminum metal, performed in a liquid cryolite bath in which it is progressively dissolved to alumina. This method is also known as Hall-Heroul process.

The grade of obtained purity depends on the degree of refining of the reduction of the mineral.

Among other properties of aluminum one can mention its:

- good corrosion resistance, thanks to the ability for a natural oxide layer formation;
- excellent formability due to its CFC crystalline structure;
- good thermal conductivity.

The weak points of pure aluminum are related to its mechanical characteristics, which are too modest to make a possible replacement of ferrous alloys. For this reason, the interest is mainly moved towards aluminum alloys, which contain additional alloying elements making the structure more resistant from mechanical point of view.

In the last decades, the interest toward the aluminum alloys has grown considerably, especially concerning some applications in automotive area. Al alloys reveal high strength /weight ratio, good formability, excellent combination of castability and mechanical properties, which associated to an excellent corrosion resistance, make them very exciting for a great range of applications. These aspects are directly correlated to a substantial savings in terms of weight, and then fuel consumption and pollutant emissions. In particular, it is appropriate to highlight not only the increase in the use of aluminum alloys, but also the increasing production of secondary alloys, e.g. deriving from the recycling of aluminum scraps. This process leads to further advantages both concerning the saving of raw materials, as well as from environmental point of view.

The production of Al alloys goes through a complex operation process of alloying of the various elements to the molten aluminum bath. This process does not directly dependent on the melting temperature of the elements added to Al, therefore it is possible to add metals such as Fe, Cu, which have higher melting temperatures than Al. On the other hand, the increase of the temperature rises the kinetics of dissolution. The direct introduction of the different alloying elements in the Al bath is, however, carried out only in the case where the melting temperature of the element is equal to or less than that of the aluminum, as is the case for Mg, Zn and Sn, or in the particular case of Si, which can be added directly even if it has a much higher melting temperature than that of the Al. In other cases, generally, the use of special intermediate alloys, the so-called master alloys take place.

Aluminum alloys can be classified into two main groups:

- foundry Alloys
- alloys obtained by plastic deformation

Within each of these two categories, one can talk about the heat treatable alloys and about the non-heat treatable alloys.

The alloys originating from plastic deformation are monophasic alloys, constituted by the single solid solution of metal atoms added to the alloy with those of Al; this situation occurs when the added elements are situated in a low proportion, as indicated by the corresponding state diagrams. The presence of only one phase is an indispensable condition to precisely facilitate the plastic deformation, which for Al-based alloys is usually performed in a cold way. The best condition of deformability has been obtained for pure Al. There are different strengthening mechanisms which can contribute to increase the mechanical properties of Al-based alloys:

- Work hardening, which is commonly used for all Al-based alloys as a result of plastic deformation;
- Aging, typical only of the heat treatable Al-based alloys.

The alloys obtained by foundry technologies differ from those obtained by plastic deformation as concerns the higher content of alloying elements, which further reduces the melting temperature and it can be considered as a positive aspect of such alloys, because it reduces the effort needed to obtain the liquid state. On the other hand, the monophasic field disappears causing the strong decreases of their deformability. Another feature that makes Al alloys a classic example of foundry alloys is their good fluidity, indispensable especially in the case of the production of castings with complex shape.

The element usually used to increase the fluidity property of the alloy is Si: when 11.7% (weight) is added to the alloy leads obtaining with Al at 577° C the eutectic phase. The fluidity increases with the Si content in the whole range of hypoeutectic alloys, till to the eutectic composition. At the same time, Si contributes to decrease the solidification range, with the development of a lower shrinkage and the presence of lower thermal stress on the die. This fact implies, however, a limited casting time and a more difficult filling of the shape before the beginning of the solidification phase. For this reason, the best compromise as gravity casing concerns is achieved with the use of about 7% of Si in the composition of the alloy.

Silicon positively influences also the mechanical properties of the alloy. However, it is important to take into consideration the hardness and the wear resistance of the alloy and to find a right compromise between such characteristics to avoid any difficulties related to the mechanical workings on the castings. Si in correspondence with the eutectic temperature, tends to solidify in the form of polygonal plate, which negatively affects the resistance of the tool, consuming it earlier during the machining process, causing its breaking. To avoid such difficulties, it is common practice in the foundry to perform chemical modification of the eutectic, adding to the alloy small amounts (ppm) of elements, e.g., Sr, Na, Sb able to modify the morphology of the eutectic Si particles, and in particular changing it from lamellar shape to fibrous one and developing their uniform distribution which is no longer concentrated along preferential directions. This precaution leads obtaining better mechanical properties and superior workability of the casting, reducing the wear phenomenon on the tools. In addition, there is a decrease of the eutectic temperature too. Silicon is another element of great importance in such alloys, forming an intermetallic, Al2Cu, phase with Al, which can be considered responsible for the increase of the mechanical properties of the alloy after heat treatment. A similar function has the Mg, which forms compounds with the Si, such as Mg2Si particles.

Iron is a harmful element to Al alloys, especially when the solidification is slow, such as in gravity casting process. The development of intermetallic particles of elongated morphology and acicular shape determine a brittle character to the alloy. The addition of Mn can overcome this problem by changing the shape of these intermetallic compounds modifying the shape of such particles into the so-called Chinese script in structure, which are less harmful for the stiffness properties.

The mechanical properties are also affected by the grain size of the crystals: fine-grained microstructure lead to superior mechanical properties. For this reason, it is possible to modify the alloy composition with addition of a small amount of Ti and/or B, usually added to the liquid metal bath in a form of tablets or rods of AlTi5B, determining the development of TiB in the form of extremely fine particles, acting as sites for a heterogeneous nucleation of Al, which then solidifies in a more homogeneous crystalline structure with finer grains.

In addition to the chemical modification, the microstructure of the casting is intensely influenced by the rate of solidification. Higher solidification rates determine finer crystal grains, smaller precipitates which are well distributed within the structure. A shortest interdendritic distance is obtained, remarkable by measuring the Secondary Dendritic Arm Spacing (SDAS) value.

The chemical composition and the mechanical characteristics of the Al foundry alloys are described in the UNI EN 1676, for the ingots used for re-melting, and in the UNI EN 1706 for the casting alloys.

3.1. Foundry processes

In this category of techniques are found all the methods which allow obtaining the castings in the finished product form or semi-finished product with more or less complex shape, by melting and casting of ingots of molten metal into molds. The commonly used different foundry techniques are distinguished depending on the material which made up the die and through the way in which the introduction of the liquid metal takes place. These variables have an effect on the quality of the casted shape and on the overall cost related to the whole production process.

The choice of the most suitable casting technique to be used in a specific situation has to take into consideration different simple concerns, such as:

- the required mechanical properties of the casting;
- dimensional accuracy;
- surface finishing properties;
- geometric complexity of the pieces produced;
- economic aspects.

In all foundry processes, the design and the manufacture of the dies and the models are of primary importance. The liquid metal is introduced into the molds, or the cavities which reproduce in negative the particular shape to be obtained. These can be:

- short-lived forms (to be lose), mainly made of silica sand and usable for the production of a single casting;
- permanent forms, typically made of steel obtained by different machining methods and usable for the production of thousands of castings.

In the case of the die casting process, one has to consider the mold, when high pressures are in use.

The model reproduces the shape of the pieces projected, but not its final dimensions, since it is necessary to take into account the shrinkage of the metal during solidification. The models are used to produce the forms and can be divided into:

- one-use models, made up by wax or polymeric material;
- permanent models, made up by wood or metallic alloy and used to produce forms to be lose.

Finally, in the case when the casting shows cavities, insertion, in the form, of cores produced in sand can be carried out. These elements, have to possess, from one hand, the mechanical properties required to withstand the pressure applied by the liquid metal, and from the other hand, they have to be sufficiently permeable to allow the outflow of the air from the cavity and to facilitate the extraction of the solidified casting.

3.2. Considerations on the presence of defects in casting

Shrinkage porosity is one of the most important quality problems directly connected to the casted products. It is developed when cast material takes up less space when solid than when liquid, and this space appear where there is a hot spot in the casting. The way to reduce or eliminate the defect is by increasing the pressure on the semi-solid metal at the porosity location during solidification by hydraulically actuated action. However, the regulation of the pressure is often difficult, due to the variations on the metal and die temperature. There is a necessity to control the employed parameters, to have a stable process, acting on the accurate temperature, pressure and force, because any inaccuracy coming from the control process can promote a non-quality stable product. The thermal stability of the die should be also achieved, because if there are hot spots in the die, the shrinkage porosity can appear easily.

When a non-quality part is detected in the process or it will be detected by the final customer, only reduced information about the part processing can be determined. This lead to the augmentation of quality control and cost. The project will develop processes which collect and store the information's in order to increase the part traceability.

The quality of the component is strictly correlated to the thermal control of the die during the process. In fact, the presence of internal defects in the cast, its final surface quality, the life of the die itself, can strongly depend on the temperature distribution of the whole considered system. The introduction of a cooling system that considers the chemical composition of the alloy and therefore of the final component, its shape and its final mechanical characteristics required, as well as all the parameters involved in the casting process, determines an effective quality change and obviously an increase of productivity.

During the production process, the liquid has high tendency to shrink during solidification, for which the design of the sprues has to be performed adequately.

Additionally, Al has also strong tendency to form, on the metal bath, a surface oxide layer, which can break during the casting operations and in this way it can be considered a source of the development of non-metallic inclusions (hard points in the casting). Their presence would be a disadvantage to the quality of the casting and can reduce the life-time of the molds and tools which are used for the machining of the parts. For this reasons, they have to be absolutely avoided by performing a correct action of the hot liquid metal bath and avoiding the turbulence of the flow during casting.

Another main source of defects of Al alloys is the gas porosity, due to the solubility of hydrogen in Al, which is high in the liquid state, but practically inexistent in the solid state. During the solidification, the hydrogen is then excluded from the liquid metal alloy in the form of porosity of a round shape, easily distinguishable from the shrinkage porosity. The presence of these defects can be avoided through a correct casting with laminar flow and by appropriate degassing treatment, performed in the ladle by blowing N_2 or Ar; these inert gases usually create an accurate agitation of the bath and can promote the outflow of H_2 bubbles.

Finally, it is very important to control the (i) temperature of the liquid metal bath, in order to avoid the loss by volatilization of some alloying elements, and (ii) the temperature of the shells or the molds, in order to avoid any thermal shock.

Casting in permanent molds leads to the production of high quality components. The metal walls of the shell or of the mold in fact favor a high cooling rate, and then a fine microstructure that allows achieving high mechanical properties. At the same time, a higher surface finishing properties and higher dimensional accuracy have been obtained compared to techniques using forms that are losing.

The choice of permanent mold casting implies an important investment as for the plant concerns and usually it is justified only by a large-scale production. The most important methods using permanent forms are:

- gravity chill casting;
- low-pressure die casting;
- diecasting.

To these already well-known and widely used techniques at an industrial level, addition of some innovative processes can be considered, where they exploit different approaches concerning the conditions in which the alloys are processed or as regards the engineering plant employed, which all together perform an improvement of the quality of castings and at the same time by reducing the development of scraps and thus the production costs.

The main techniques are, for example, the rheocasting and thixocasting processes, that use some gains coming from a semi-solid approach, where the liquid metallic alloy is injected into the mold in a defined temperature range, corresponding to the interval of the solidification.

Another methodology is the squeeze casting technique, which can be described as a forging of the metal in the liquid state.

As an alternative one can use the Fondarex or Vacural processes, exploiting the technology of the diecasting and it is performed in vacuum. In the first case the vacuum is created only in the mold cavity while in the second case in the whole plant.

Gravity chill casting: the casting is carried out in the source, and then the liquid metal flows along the casting channel and goes to fill the mold from the bottom through the ingate, arranged in number and in an appropriate position to guarantee a suitable filling of the form and of sprues or risers. The flow rate has to be such to allow that the metal reach every point of the shape before the beginning of the solidification and without creating turbulence.

The steel walls of the form that come into contact with the liquid Al are normally coated with a ceramic layer, with the purpose of, from one hand, to limit the wear of the implant and from other side, to reduce the solidification rate, limiting the heat exchange. In this way, choosing an adequate coating thickness, it is possible to adjust the liquid metal solidification rate and to distinguish between different zones in order to obtain the preferred microstructure.

The presence of a metallic or ceramic filter at a certain point of the casting channel is sometimes justified by the need to suppress the impurities present in the metallic bath.

After the solidification, the metallic piece is extracted from the shell, constituted by at least two parts, to allow accurately the demolding of the piece. The following operation step is related to the mold assembly, i.e., the positioning of the cores and of the shell which is closed again for the next casting. This whole process can be automated and the velocity can be enhanced by placing more shells on the carousel, which rotates at programmed intervals. In this way, during the solidification of the liquid metal in the form, it is possible to process other castings.

A significant part of the production plant for casting is designed for the production of cores, which plays a fundamental role in order to obtain high quality castings. An incorrect production of cores or an inaccurate positioning inside the shell constitutes a further source of defects for the castings in the shell, in addition to the classic defects present in the pieces for the foundry products, e.g., gas or shrinkage porosity, the presence of hard points and cold junctions in the casting. The last mentioned features are in particular points where the metal is not joined correctly because two flows came in contact at a too low temperature when the parts are already partially solidified. This situation, which usually occurs in more distant areas from the casting attacks, favors the development of micro-cracks which drastically reduces the mechanical performance of the alloy and in particular the sealing pressure of the castings.

Low-pressure die casting constitutes a variant of the previous method with regard to the filling system. In this case, in fact, the metal goes back in the form following the application of an external pressure on the crucible, applied so as to maintain the laminar flow and a filling as much as possible regular.

In addition, the pressure also guarantees the supply of the metal during the solidification phase, thus reducing the problems related to the shrinkage without the need to use additional metal investments.

The modifications performed to the casting process allow obtaining improved quality castings, with the following benefits:

- reduction of impurities in the bath;
- finer microstructure and enhanced mechanical properties thanks to a higher solidification rate;
- reduction of the scraps, since there are no sprues and the runners is of small size.

Compared to gravity casting process, such type of system involves major investment in terms of cost and space, because its higher dimension.

3.3. Heat treatments for Al alloys

The mechanical behavior of materials is influenced by their chemical nature and the mechanical and thermal processes to which they are subjected. In particular, any heat treatment induces a modification of the crystal grain size, on the diffusion and coalescence of the porosity and on the presence of precipitates within the casting.

Generally, Al based alloys are subject to two other forms of hardening, in addition to the possibility to the formation of precipitates. In the first place they undergo to hardening by solid solution, due to contribution of the crystal lattice distortion generated by substitutional alloying elements in substitutional position. The alloys from plastic deformation are also subjected to strain hardening, a permanent change in their form.

Mechanism of precipitation hardening is typical of those alloys in which the solubility of one metal in the other one decreases with temperature, leading to the expulsion, during cooling, of the intermetallic compounds from the matrix. As Al alloys concern, this behavior is typical of the Al-Cu systems, Al-Si-Mg, Al-Si-Cu-Mg and Al-Zn -Si.

The alloy is firstly homogenized for solubilization by heating in the monophasic region for the necessary time in order to solubilize all remaining soluble particles, and then it performs a rapid cooling to a suitable temperature in the two-phase region. The rapidity of this cooling prevents the formation of precipitates at the equilibrium state and thus produces a supersaturated solid solution. During ageing at room temperature or slightly above, development of transition fine structures occurs with the size of about 10 nm.

The hardening mechanism by precipitation during heat treatment consists of three stages, and in particular annealing, quenching and ageing. Some details will be discussed with the focus on Al-Si-Cu-Mg alloys. The heat treatments used in the production of Al alloy castings differ from each other due to the presence of the solubilization and the way in which ageing is carried out. The most important heat treatments are as follows:

- T1: solidification and natural ageing followed by controlled cooling;
- T4: solution heat treatment and natural ageing;
- T5: solidification and artificial ageing followed by controlled cooling;
- T6: solution heat treatment and artificial ageing at T < 200 ° C;
- T7: solution heat treatment and artificial ageing at T > 200 ° C.

Within the above mentioned thermal treatment T6 treatment is the mostly industrially, even if for some applications the research is moving towards the T7 treatment.

The solubilization phase has to be performed at a relatively high temperature, close to the eutectic temperature of the considered alloy. The aims of this treatment are to:

- dissolve the soluble phases containing Cu and Mg, formed during solidification;

- homogenize the contents of these elements in the Al matrix;
- modify the shape of the eutectic Si particles from a lamellar structure to a globular one, with positive effect on the mechanical properties.

All these processes have diffusive nature and consequently are thermally activated, the rapidity of the evolution increases as temperature increases. This parameter also affects the hardness of the alloy which is obtained at the end of the aging treatment and depends on the amount of elements that pass into solid solution during the solubilisation phase. The maximum temperature of this phase of treatment is limited by the melting temperature of the low-melting point eutectic phases. The chemical composition of the system considered has to be taken into account for a correct setting of the heat treatment parameters.

In the case of the AlSi7Cu3Mg alloy with a Mg contents between 0.3 and 0.6% (wt), the presence of this element causes segregation during solidification of phases such as β-Mg2Si and Q-Al5Mg8Si6Cu2, in addition to the Al2Cu particles. The presence of Cu compounds, in particular of the Q-Al5Mg8Si6Cu2, limits the solubilisation temperature for these alloys to below 500 °C, due to their initial melting which occurs when one remains at 505 °C for a sufficiently long time. Higher temperature would lead to the fusion of these compounds with following solidification without any crystalline structure.

The chemical composition also affects the tendency to dissolve the compounds formed in the as cast alloy. Some studies report different behaviours of the Q-Al5Mg8Si6Cu2 phase as a function of the Cu and Mg percentage and of the variation of the temperature: sometimes these particles tend to dissolve, sometimes remain stable and occasionally increase negatively affecting the β-Mg2Si phase. Furthermore, not all phases are soluble, e.g., Cu or Mg atoms within the intermetallic compounds with Fe will not pass into solution and will not contribute to hardening the alloy.

When the heat treatment is setting, one can take into consideration also the production process of the castings, because the manufacturing route determines the preliminary microstructure of the alloy, and then the necessary time to guarantee the correct diffusion and homogenization of the elements which dissolve in the solution. A rapidly cooled alloy has a finer microstructure, as well as a modified or refined alloy too; the dissolution time will be lower and therefore an inadequate extra time would damage the growth of the crystal grain.

Recently, various studies have been focalized on the thermal treatments of Al alloys in order to propose alternative solutions as time and temperature concerns and to obtain higher mechanical characteristics as well as an optimized industrial process. A possible variation, commonly performed industrially, involves heating in a single step and to carry out the treatment in two stages characterized by different temperatures and times. In the first stage the alloy is solubilized at low temperature, 495 °C for 8 h, as in the case of single-stage. In these conditions the dissolution of the Cu containing phases occurs without any risk for the melting, but the kinetics of the diffusion is limited and the modification of the eutectic Si results to be difficult. In the second stage the temperature is raised to 520 °C for 2 h, to favour the homogenization and to guarantee the conversion of eutectic Si grains into spheroids. As results, one have an improvement in both mechanical strength and ductility. The only drawback is related to the considerable increase of the solubilisation time, which results approximately doubled compared to the traditional treatment.

The aim of the quenching is to move the structure formed during solubilisation to room temperature. Through rapid cooling a supersaturated solution is obtained (which is only possible by means of a high cooling rate) otherwise the precipitation of the elements at the grain boundaries occurs, with their reduction in the supersaturated solution causing a poorer increase of the mechanical properties due to ageing. However, quenching induces strong thermal stress on the casting and high residual stresses, which can be reflected by the distortion of the pieces, lower dimensional accuracy, development of cracks due to the stress created between the edge and the core of the pieces.

Foundry alloys are more inclined to quenching compared to alloys produced by plastic deformation, due to the presence of the eutectic Si. A very rapid quenching maintains the Si atoms in solid solution with Al and limits its diffusion to the lamellar Si eutectic phases formed during solidification; these atoms will then be available to form Mg_2Si during ageing. A high density of dislocations is due to the presence of the eutectic Si because of the different thermal expansion coefficient of Si compared to Al, which acts as nucleation sites for the precipitates. All these features have an impact on the mechanical properties: as cooling rate increases the yield strength increases.

It comes out that the choice of the medium of the quenching is of great importance; a simple cooling in air in fact is not enough to freeze the supersaturated solid solution up to room temperature. Currently, water is the most widely used industrially quenching medium, since it allows a very rapid cooling. To reduce the drastic nature of this treatment it is possible to use water heated up to about 75 °C, thus reducing the ΔT, or by applying a quenching delay, between the solubilisation and quenching in water, paying attention on how this should influence the final mechanical characteristics of the casting.

In some situations, when it is necessary to further reduce the velocity of the treatment, this is the case of pieces which show high risk of cracking because their sharp edges, it is possible to use quenching in oils, salts bath or organic solutions. In such cases, their lower thermal conductivity compared to water is a benefit, reducing the heat exchange of the casted piece. The use of these substances requires additional attention for monitoring the process to avoid the release into the environment of harmful substances.

A specific quenching, providing a cycle divided into three stages, as follows:

- First stage: quenching in boiling water (T =100 °C, t= 15 min);
- According to the quenching stage in liquid N_2 (T = -196 °C, t= 30 min);
- Third stage in boiling water (T= 100 °C, t= 15 min).

It has the advantage to greatly reduce the residual stresses in the casted piece, and then the risk of sprains, without causing a variation of the microstructure and a reduction in mechanical properties. This is possible because it reduces the ΔT at different stages and because the last stage of rapid heating induces in the casting opposite residual stress to those caused previously, which therefore has been neutralized. As disadvantages one can mention an enlarged timeframe that usually takes a few minutes and a higher cost, due to the use of N_2.

The supersaturated solid solution obtained by hardening is not thermodynamically stable and over time it will evolve towards an equilibrium situation, creating a distribution of precipitates uniformly distributed in the Al matrix. The phenomenon that leads to precipitation of Al_2Cu and Mg_2Si phases it is performed by ageing. One can talk on natural ageing when the ageing takes place at room temperature in an extended time, artificial ageing, when it carried out in oven at a temperature between 150-220 °C for several hours. In particular, when the aging takes place at T <200 °C i.e., performing a T6 treatment, while when it is pushed at T> 200 °C a T7 treatment is carried out.

Increase of the hardness and of the mechanical resistance, due to ageing, is given by the capacity of the precipitates to delay the movement of the dislocations and therefore the plastic deformation of the alloy. The interaction between dislocations and precipitates can be described according to the Friedel effect and to the Orowan mechanism. The first describes the sharing effect that the dislocations apply on small precipitates with low hardness; the second one refers to some hard and larger in size precipitates, which are bypassed by the dislocations. The resistance of precipitates increases with the size up to which the dislocations are able to share them. Additional dimensional growths are harmful on the development of the precipitates governed by the sharing mechanism and it leads to the activation of the Orowan mechanism, according to which there is a progressive decrease of the resistance of the precipitates as their size grows.

The highest level of resistance is achieved when the probability that the dislocations have to bypass or share the precipitates is equal.

The final properties of the treated piece are governed by the mode in which the ageing is carried out, by the chemical composition of the alloy and by the thermal history of the alloy that in turn is affected by the type of precipitates developed.

Adding small amounts of Mg to Al-Si-Cu alloy, one can obtain higher mechanical strength than the ternary alloy, but lower elongation at break. Talking about Al-Si-Mg alloy the quaternary alloy shows slightly superior properties, but there is an increased time to reach such properties.

Comparing three alloys, Al-Si-Cu-Mg alloy shows the best response to the heat treatment with respect to the other two taken into consideration, due to the precipitation of Al_2Cu and Mg_2Si phases. The latter one guarantees superior mechanical properties compared to Q-$Al_5Mg_8Si_6Cu_2$ phase, developed in a higher percentage during the high temperature treatment, motivating the lowest yield strength. Moreover, the same Mg_2Si phase at above of 200 °C, precipitates in the form of β' and not as β'', responsible of the softening (as explained also by the hardening mechanism by precipitation) which lowers the resistance of the alloy. The ageing curves related to the yield strength and to the tensile strength of Al-Si-Cu-Mg alloy, highlights the influence of the microstructure, measured by the Secondary Dendrite Arm Spacing index; the microstructure being equal, it evident that T7 heat treatment, with aging temperature higher than 200 °C, allows obtaining higher mechanical properties earlier, even if these values are slightly lower than those obtained with T6 treatment, with lower than 200 °C ageing temperature.

The elongation at break shows a more complex behavior, since it is intensely influenced by the microstructure. In a fine microstructure, in fact, the eutectic Si particles are more regularly distributed and therefore they are able to delay the movement of the dislocations. These overstep the Si particles as a result of a shearing effect and they are accumulated on the grain boundaries, where the fracture is originated, thus allowing a superior elongation. Contrarily, in the case of a coarser microstructure, the dislocations are accumulated near to the Si particles forming a continuous barrier; in such cases, an intra-granular fracture takes place with a lower elongation.

The effect of the microstructure is more accentuated in the alloys which contain Cu, where the intermetallic particles developed with Fe with elongated shape are insoluble during the heat treatment and become much more brittle compared to the smaller (β-Fe) phases, a typical development in Al-Si-Mg alloys.

The treatment affects the elongation since it modifies the morphology of the eutectic Si. The transition to an ageing treatment at higher temperature (T7) the elongation slightly increases due to the fact that the values for the strength at break are lower compared to the case of a T6 heat treatment.

Finally, in the scheme of the heat treatment it has to be taken into consideration that a possible natural ageing prior to artificial ageing can have a negative effect on the performance of the material. During the permanence at room temperature, precipitation of clusters atom occurs and if the radius of these particles is larger than a certain critical radius, which increases with temperature due to the minor oversaturation, they become stable and can act as nucleation site during the following artificial aging. However, if the dimensions are smaller, the clusters become unstable and they can dissolve in the solid solution by increasing the concentration of the solute in the matrix; in this way the critical threshold decreases and other clusters become more stable. This negative effect can be eliminated by a short high-temperature step before artificial ageing or by a short ageing carried out immediately after quenching.

3.4. T6 and T7 heat treatment of Al-Si-Cu-Mg alloys

The difference between the above mentioned treatments is the temperature at which the artificial aging is carrying out. Forcing the temperature above 200 °C leads reaching higher mechanical properties in a shorter time since the formation of the precipitates is a thermally activated diffusive phenomenon.

4. Die-casting

Die-casting, a more recent technique compared to gravity casting, involves the injection of the liquid metal into a mold by the application of a high pressure. The process initially involves the filling of the supply channel with the necessary quantity of molten metal, which is then injected inside the mold by a piston which slides along the feed channel. In this step one can highlight three phases, characterized by different velocities and different pressures:

- first phase in which the metal is pushed up to the attack of casting. This stage is characterized by velocity of the alloy flux and low pressure, in order to minimize the turbulence of the flow, the most important cause of scraps due to the presence of porosity;

- second stage of mold filling. The piston applies a higher pressure and the liquid metal flows into the cavity of the mold at high speed. The mold has to provide appropriate rushes and wells for the evacuation of the air and of the excess of lubricant, which is more difficult due to the high filling rate;

- third stage (step of multiplication of the pressure). The mold is filled and the high pressure applied helps to feed the liquid metal required to balance the solidification shrinkage.

After the solidification, the mold, consisting of a fixed die and a movable one, opens to allow the elimination of the piece. A lubricant material is used on the inner walls of the mold which is necessary to facilitate the extraction of the part and to limit the wear of the mold, principally caused by the chemical aggressiveness that Al has in contact with steel, since the molds for diecasting are used with no any coating.

This process guarantees to obtain good quality castings, with very fine microstructure developed with high solidification rate. The alloys used have a significantly higher Si content than that for gravity casting process in order to maximize the fluidity of the alloy. Very thin castings can be produced, with the drawback of no possible internal cavities, since the high pressures applied exclude the use of cores.

As already mentioned above, the main problem of die castings is the development of gas porosity, which generally affects unfavourably the mechanical properties of the pieces.

The cost of this process is very high and frequently it is justified only by a large-scale production.

Sand casting, lost wax casting (Lost Wax) and lost foam casting (Lost Foam) are the most important casting procedures.

Sand casting substantially uses the same scheme of the permanent mold gravity casting, where, however, the shape is no longer metallic, but made up precisely by a mixture of sand silica, water and bentonite, which acts as a binder and imparts to the sand the necessary plasticity and cohesion properties. In addition to the mixture above mentioned some additives are used, which are useful to improve the refractoriness of the sand. The following properties are required:

- plasticity, to loyally reproduce the model profile;
- cohesion, to retain its shape even under the pressure applied by the liquid metal;
- refractoriness, to withstand the thermal stresses;

- permeability, to allow the ejecting of air contained in the form and of the of water and steam developed in the case in which the material used is humid.

The molding is carried out for pressing the sand on the pattern plate inside the bracket, a metal structure that contains the shape in the sand and allow to resist the pressure applied by the liquid metal during casting. Usually the forms are divided into a lower half and an upper part; the latter also contains the runner and the sprue.

Once the sand is compressed, and possibly dried, in case of using a dry forming rather than one green, the two parts are assembled and then casting could star. When the solidification has finished extraction of the casting takes place.

Alternatively, it is possible to create shapes in sand, which are able to resist the pushing of the liquid metal even without the brackets, simply supporting each other if arranged in series. The resulting advantage is a considerable reduction in the number of brackets and then the space occupied inside the plant.

Compared to permanent mold casting this process requires less investments, but leads to the production of pieces of inferior quality, since the solidification of the liquid metal takes place in sand, with much lower thermal conductivity compared to a metallic mold. For this reason, it is possible to use some metal inserts inside the sand form in correspondence of critical points of the casting to which superior mechanical characteristics are requested.

The lost wax casting is a very old process, but far from simple. It is used to produce parts in small series of reduced sizes and shapes, even in a very complex character, since it guarantees a very good dimensional accuracy and surface finishing properties.

The process consists in realizing the model of the pieces in wax, in order to reproduce with high reliability, the shape and the size of the piece to be produced. More models are assembled along a sprue also made of wax to form the so-called cluster, which is then immersed in a ceramic solution typically gypsum powder or very fine-grained silicates, to form a shell that guarantees a high surface finishing properties. The immersion is then repeated several times in a solution containing powders with increasing particle size, in order to let the shell reaches a necessary mechanical characteristic. During the drying process the wax melts and flows out from the shell and in the meantime it consolidates. The metal is casted within the ceramic form obtained.

The main advantages of this technique are limited to a necessary finishing machining and the possibility of being able to produce more pieces with a single casting. On the other hand, it is necessary to pay attention to the implementation of the shell, in particular on the size of the powders used.

The lost foam process represents an evolution of the casting in lost wax casting, in which the model is produced in polystyrene, also in multiple parts if the workpiece geometry is very complex. The models are then made up by gluing the various parts and assembled together with the sprue to form a cluster that is subsequently painted by immersion in a ceramic solution.

The substantial difference is related to the fact that the model is not removed from the shell before casting, but it burns during the casting itself. This, on the one hand, eliminates the need of cores or the difficulty of the extraction of the model from the shape, on the other hand, however, some problems arise with a correct elimination of the gases deriving from the pyrolysis of polystyrene and to avoid the presence of porosity in the castings. To overcome this issue, the coating layer has to be sufficiently permeable to the gases.

After drying, the clusters are placed in containers which are then filled with sand, with the function of supporting the shell during casting. In this case a total recycling of the sand used is possible, since it is not polluted with other substances, but only carries out its function of support in a form of a simple compressed powder around the shell.

The main advantage of the casting process in lost foam consists in being able to produce even very complex parts with a good surface precision and the almost total absence of burrs, thus reducing to a minimum the necessary finishing operations.

The presence of porosity, due to the gases produced by the combustion of the model or to the difficulty of the supply of the metal due to the presence of a single sprue, remains the main problem of this technique to produce high quality pieces.

4.1. Some consideration on the use of dies for manufacturing components

Dies and molds show a minor investment with respect to the total cost of the complete manufacturing platform. Nevertheless, they are significant such as other part of the engine, because they are directly or indirectly correlated to the length of the process, quality and expenses related to the produced components. Additionally, the quality of the die in a straight line marks the feature of the casted piece.

The die is the element that directly provides the shape of the casting. The precision and the high quality of die casted products strongly depend on the performances of the die, as well as from its correct use on the machine. The die is constituted by a set of elements built and assembled with precision, and its size significantly depends on the shape and the casting dimension to be achieved, as well on the mechanical stress, thermal and wear mechanism to which it is subjected during operation. All these aspects are of considerable importance for the service life of the die and for the volume of castings which one propose to produce. The assembly of the elements constituting the die has to be realized, in the closed position (i.e. corresponding to the condition when the die is closed), a profile of the cavity with a shape and a size of the casting to be produced and of the various elements which made up the supply system where the liquid metal is injected. The die has to show several requirements: the most important can be summarized as follows:

- to guarantee, at the operating temperature, the required movements of the movable parts;
- to withstand the thermomechanical stress due to the casting/injection of the liquid metal;
- to guarantee, after solidification of the metal, the expulsion of the casting without ruptures and deformations.

In its simplest form the die is made up of two main parts: a fixed one, attached to the fixed plate of the machine, and a movable part, situated on the movable part of the machine. On the movable part, and sometimes even on the fixed one, they can be of the radial rods and their presence depends on the particular shape of the pieces to be casted or on the need to create holes that are not overlapped by the plane of division between the two half-dies.

The realization of permanent dies for the manufacturing of parts on industrial scale shows numerous challenges, which are different from casting to casting because all the process variables are characteristics for each singular production. Therefore, in order to obtain the maximum functionality and productivity, it is necessary to take into account many requirements and numerous critical aspects of different materials used for the realization of the dies. Actually, the development of new materials is not yet able to replace totally Fe-based materials (steels) traditionally used in the fields of mechanical processing, in the forming by plastic deformation and, specifically, in the field of diecasting process. The research in this field is constantly growing, and mainly associated to the fact that the steels normally used are highly stressed by the extreme operative conditions, and because they arrive premature at the end of their life, involving higher production time, increasing the production cost, etc.

Although the plan signifies only a lesser portion of the complete fabrication, choices completed at this point have an effect on the development and on the life cycle costs of the casted component.

Additionally, to practical necessities, the part design has to contemplate the restrictions of the production process and the tooling requests, tolerances and fabrication rates and finally the characteristics of the alloy which will be managed using the selected die.

Actually, the production phase of a die is preceded by a long stage of modelling and design with finite element based software. It is believed that the role of the process simulation in any foundry technologies (but also in other technological fields), is a crucial and critical point. There is no doubt that these tools make it possible to modify both the production technique and the design parameters, but it is equally important, however, to be aware that the direct experience in this specific field cannot be replaced by any software, no matter how expensive, sophisticated and complete it can be.

The currently CAD and CAE systems available on the market allow to drastically reduce the industrialization of the product times, ensuring excellent results, thanks to the optimization of the process parameters as a function of the casting functional characteristics, combining the full potential of the press and the characteristics intrinsic properties of the alloy used for the casting. In fact, the numerical simulations simplify the understanding of the relationships between the different elements that contribute to the design of the die and the various parameters which are of fundamental importance for the aim of the process prediction, the necessary production condition of quality pieces and constant submission. The geometric definition of the die and the prediction of the thermo-mechanical stresses, which act on the die, in relation to the operating parameters of the tools, represent the set of results that can be verified at the design stage by means of fluid dynamics and thermal analysis following all the phases of the production cycle of a component. Therefore, the simulation of the filling and solidification steps, and of the thermal cycles to which is subjected the die, allows to evaluate all parameters which are fundamental during the life of the die service, such as the material flow and the induced stresses, adjust process variables, such as temperature and the holding time, design constructive and structural parts in an optimal manner, and eliminate the formation of possible defects in the castings produced. By simulation methods it is possible to predict the solidification structures too.

As part of the manufacturing cycle of steels for dies and for tools are normally starts from a selected scrap. The first phase of the process takes place in the electric arc furnace with successive refining and vacuum degassing in the ladle furnace; then polygonal ingots are reached and they have to be distributed in the forge department or the ingots have to be re-melted, in the electro-conductive slag (ESR) or in vacuum (VMR), to produce new forging ingots. This last operation, carried out with special devices, in order to give to the product practically equal mechanical properties in all directions and thus to guarantee a constant quality over time and the chosen properties.

On the obtained bars some machining operations are performed, obtaining the cavity of the shape of the matrix. Actually, there is a technology growing up recently, for the production of dies, based on high-speed machining (HSM-with a rotation speed comprised between 10.000 and 100.000 rpm). The benefits arising from the use of such high-speed machining are multiple, including: better surface finishing properties on the component obtained (normally the roughness is less than 0.3 µm), higher amount of removed metal, a significant reduction in the processing time (an average of 30%) and consequently reduction of the production cost, guarantee to develop geometries with very thin walls, since the shearing forces and vibrations are significantly reduced; additionally there is less distortion of the material matrix, because the heat is previously removed and it is therefore possible to go on without overheating them (assuring lower deformations). The construction of the die can be summarized in the following stages:

- shaping/profile treatment (milling);
- heat treatment in order to obtain the desired hardness;

- rapid milling;
- grinding and manual finishing.

After the shaping step and before the rapid milling, generally a thermal treatment is carried out to optimize the microstructure. This is determined by the hardening temperature and the holding time, as well as on the cooling rate and by the tempering. After the hardening/tempering, one obtains a microstructure, which corresponds to a hardness of 48 HRC maximum; higher values can lead to rupture during use.

Following heat treatment, the previously rough-machined die is subjected to a high-speed machining by precision machines from three to five simultaneous axes. The linear motor technology in all five axes allows to obtain, together with an absolute stability, minimum processing times, excellent quality of the machined surfaces with Ra < 0.2 µm, high imperfection removal per unit of time, and it is useful also for the construction of complex dies.

With an additional final treatment (grinding and polishing) the surface structure is improved and also the cracks are eliminated up to a depth of 5-19 µm.

4.2. Thermal treatments of steels for dies

The steels for hot working tools are generally supplied in the annealed condition. After the creation of the cavities, the die has to be subjected to a thermal treatment (hardening) in order to obtain the optimum values of resistance (hot yield strength, tempering resistance, toughness values, hardness and also the desired ductility). A steel which has been subjected to an incorrect heat treatment, can give inferior performance, in terms of physical-mechanical properties and from structural point of view compared to those obtained with a suitably treatment.

Quenching is performed on the semi-finished die; an adequate material quantity (noticeably higher than the deformation that arises in quenching) is required. Some requirements to be taken into consideration are the geometric complexity of the die, the relationship between thicknesses and lengths, the variations in section, the finishing properties, the holes and their distances, and especially the sharp edges, which are frequently constitute the nucleation sites for cracks development, because there are considered point of concentration of tensions/stresses.

The characteristics of the heat-treated steels depend on the austenitization temperature, the holding time at the maximum temperature, the cooling rate and the tempering temperature. During the heating, some preheating is performed to reduce the temperature differences between the surface and core, thus achieving the maximum thermal equilibrium on the whole mass to be treated. They are particularly important in case of very different sections in order to obtain the best compromise temperature-maintenance time. As for the austenitizing temperature, higher values have a positive effect on the tempering resistance and on the hardness at higher temperature, reducing the tendency of the development of thermal fatigue cracks.

These properties can be improved by using austenitizing temperature of 1010-1050 °C (which exceed the transformation temperatures); however, it has to be kept in mind that high temperatures austenitizing reduce the ductility and toughness, determining the development of a high dimension grain and the appearance of carbides at the grain boundary. These phenomena can lead to the failure in particular in the case of medium and/or large size die; therefore, such austenitizing temperatures are limited to small size dies, inserts and cores. Finally, high hardness has a favorable effect on resistance to crack due to thermal fatigue, but increases the risk of fracture. The choice of temperature and permanence at the selected temperature (the latter one with lower effect) has to be a result of a compromise which is based on the chosen characteristics to be obtained.

The rate of cooling from the austenitizing temperature has a great importance for these types of steel. The CCT and TTT curves allow identifying the times and the temperatures to be observed in order to determine the cooling curves, identify critical phases, choose the criteria to arrive and/or holding at a certain temperature. The cooling medium has to guarantee an optimal heat exchange, it has to be constant and uniform all over the whole casting which has to be quenched, with a suitable volume and an adequate heat dissipation to achieve the required cooling.

The cooling has to be such as to produce the martensitic structure, to avoid as much as possible the precipitation of carbides (which inevitably occurs at the grain boundary by reducing the toughness of the material) and preserve the die from fracture and excessive deformation. It is reasonable that massive particular as die of large dimensions and very important thicknesses, especially with poor heat exchange surface, will have higher difficulty to dissipate the heat and thus to create the ideal conditions of hardening up to the core of the material. Low cooling rate offers the best dimensional stability, containing and reducing the deformations, but at the same time cause an increased risk related to the modification of the morphology. In addition, a too slow cooling rate can reduce the fracture toughness of the steel. On the other hand, high cooling rates, such as quenching in salt bath, provide the best possible structure and consequently the maximum duration of the die. It is therefore necessary to find, periodically the best compromise between the lower finishing costs (to eliminate the deformations) coming from the low cooling rate and the higher duration of the die, in relation to the characteristics and geometry of the die. In most cases, for high duration of the die, high cooling rate is preferable.

The modern quenching furnaces operate under vacuum or in a controlled atmosphere, in order to avoid that the heating at the austenitizing temperatures causes the decarburization of steel (loss of part of the carbon), determining an early cracking by thermal fatigue and deterioration of the die.

After cooling at a temperature of 50-70 °C, the die has to be subjected to a quenching treatment, in order to stabilize the morphological structure of the steel and to eliminate any stresses induced during cooling. The quenching temperature is chosen according to the desired hardness value;

In order to guarantee the best possible structure, generally two tempering are performed: in special cases it can be up to five tempering. For all tool steels for hot working the corresponding secondary hardening temperature is around 500 °C. This is the temperature at which the precipitation of carbides occurs and simultaneously the residual austenite transformation (which is present in small amounts) into martensite takes place. For this, there are the needs of at least two tempering, and sometimes it is recommended also three to reach the optimum metallurgical and stable situation. A good practice provides a sequence of tempering in which the first one is carried out at 550 °C, the second one at the temperature corresponding to the desired hardness, while the third one will be performed as the first one or if necessary at a temperature which best makes possible to reach the necessary hardness. As already mentioned, high hardness has a beneficial effect on the resistance to thermal fatigue cracking, but increases the risk of fracture. In fact, the hardness to be obtained depends both on the required strength during operation, both by the choice of the mechanical machining to which the die is subsequently subjected: high hardness values of the steel accompanied by the stresses induced by the machining operations can lead to the formation of destructive internal stress states, which amplify the stresses to which the die undergoes during the operation: all these aspects lead to the early fracture of the die which are starting from the developed cracks.

4.3. Thermoregulation and mold lubrication

To achieve technological properties and constant quality of die castings it is necessary to guarantee the achievement of a complete solidification in an as possible constant time all over the casted component. This condition can be reached only having the possibility of heating the colder zones (which are mostly the thin thickness) and simultaneously cool the hottest areas (with is a larger thickness), or where, in relation to the shape of the component, there are areas of heat concentrations (areas with restricted thermal flow). Development on a die a uniform conditions, e.g., time of the solidification, is objectively difficult, but it makes feasible removing and dissipating, in the interval between one injection and the other one, the flow of heat absorbed from the die, and thus realizing the conditions related to a thermal equilibrium. This is possible by operating using a sufficient thermoregulation of the die, allowing to heat and cool each zone of the die, and operating with a good lubrication, which guarantees the thermal control of the surface.

The circuit used for the control of the temperature in the die is basically achieved by means of suitable hydraulic channels realized with the same thickness of the inserts figure and inside of them water or a heat transfer fluid is circulated. Knowing the characteristics of the heat transfer of the die, it is possible to vary the heat flux removed during the cycle by varying the flow rate or the temperature of the cooling fluid. When water is used, to achieve thermal equilibrium, commonly the flow rate is varied; in the case of oily fluids, however, modification of the temperature is preferred.

The channels for the control of the temperature realized in the die are often limited by geometrical restrictions due to the conformation of the figure, the presence of the extractors and the need not to go too close to the treated surface in order to avoid premature fracture of the piece.

The die temperature which, under normal production conditions should remain constant for each zone of the surface and at the same time in within the cycle, is actually influenced by discontinues due to different causes, but which represent the reality in the practical life in the foundry and due to the changes of some external conditions such as: non continuous capacity, irregular short-term breaks, variations of the metal temperature, and seasonal variations of the water temperature and air blustering.

The lubrication of the dies in die casting process can be considered as an important and a high level quality of the castings and a high production rate are strictly related to a proper lubrication. The volume and the type of product, its dilution and distribution on the figure of the die are important factors and often they are not considered as they merit. Actually, defects on castings, poor quality, low productivity, fragmented plugs and production breaks are often attributable to an inadequate lubrication of the figures. The duties of the lubricant are varied, and the intrinsic quality of the product has to be supported by a proper use of technique and application, requiring a good knowledge of chemical and physical mechanisms that are involved in the process.

The main functions of the lubrication of the die can be summarized as follows:

- formation of a separation layer, which prevents direct contact between the steel made up the die and the liquid metal (Al), acting as a barrier between the two, thus avoiding that Al in the liquid state dissolves Fe (according to the typical reaction $Fe + 3Al \rightarrow FeAl3$). When this barrier does not exist, or it is destroyed by the flow of the liquid alloy that arrives in the die, the result is the metallization of the die cavity and the formation of welds between alloy and die. In addition, the lubricant, as well as prevent attack by liquid aluminum, also performs a function of heat insulation, avoiding a premature solidification of the alloy, which would delay the optimal filling of the die cavity;

- formation of the lubricant film, which is formed initially as a result of spraying, after the injection of the alloy is thinned and partially destroyed and it remains nothing more than residual fractions. These have the mission of lubricating the casting in the extraction phase, favouring the detachment due to the development of gas coming from the decomposition of the film, which apply a certain pressure. The dynamic aspect of lubrication has also to be considered: namely the lubricant reduces the friction between the die and casting, favouring the sliding of the latter one on the form that has been generated, and then favouring also the extraction of the casted product. Insufficient or unsuitable lubrication, will cause deformation, cracking or fracture of the casting;

- cooling: such function of the spray is given mainly by the water which is used in the phase of the wetting of the die. The pulverization of the release agent used is made with the use of compressed air, of which action is very low.

The optimal necessity involves the use of a release agent with a low specific heat and high lubricating ability in the start-up phase of production, when the die has not yet reached the operating temperature, and the possibility to vary the spraying time, depending on the die temperature variations, when the tool is fully operational. As a result of incorrect choices of products or application techniques, lubrication can lead to different negative effects, which can be avoided and minimized, such as the formation of sticky deposits (lacquer or sludge) on the die, the obstacle of the air vents, coatings and stains on gas castings and excessive developments.

The products are marketed already mixed in various concentrations which can contain from 15-40% of active parts while the balance is constituted by the emulsion and the water. The product is further diluted for use, with water in ratios varying between 1/25 and 1/100. The emulsion thus diluted is sprayed on the die, which is typically located at a temperature between 150-350 °C. The product-lubricating and release agent can be applied manually, with a consequent variable quality of castings and an increase in the percentage of waste, or by means of automatic spraying, which eliminates the disadvantages of the manual technique thanks to the exact reproduction operation with a constant automatic cycle. The advantages of automatic spraying are: constant casting quality, minimization of the percentage of the waste and an increased production route.

The simulation models, which are actually widely used, allow making accurate simulations of the thermal dynamics of the die with different geometries of the cooling channels.

The reliability of the temperature of the die is essential for the quality of the product, for the life of the same and for the production rate.

Spraying of the release agent plays the main role in the removal of heat from the die, and thus in thermal equilibrium maintenance.

Before starting the molding of castings, it is necessary to pre-heat the die to a temperature close to that of the typical use. This treatment should be carried out with appropriate means and should last for an appropriate time proportional to the mass of the die, and has not lead in any case to tempering the less massive parts of the matrix, such as the pins and the cores.

The most widely used methods to pre-heat the die are:

- gas burners;
- equipment with infrared rays;
- hot oil circulation in special circuits or in the same channels provided for cooling.

Alternatively, one can directly inject the molten metal at low speeds and at low pressure into the die, but such a thermal shock would induce a strain on the strength of the steel, and would result in the early appearance of cracks.

The pre-heating is carried out not only to reduce the time of the start-up phase, before obtaining castings with an acceptable surface finishing properties and reduce the temperature difference between the steel of the die and the injected liquid metal, but also because this operation has a significant influence on the toughness of the steel. Thermal analysis of steels for dies is well-known: there is a transition from a ductile to brittle transition point, which lies around 150 °C; below this temperature, these steels show values that are less than half of their toughness that competes at higher temperatures. Therefore, if the die is not properly pre-heated or if such operation is performed quickly, the probability that they highlight damage to thermal fatigue is high and consequently its service life is reduced.

As concerns the maintenance of the die, in practice, it is necessary to eliminate the tensions/stresses induced by the thermal cycles. If carried out, therefore, a stress relieving at 25-30 °C below the tempering temperature, within the first 10.000 characters, and one has to repeats this operation in more or less regular intervals during the life of the die. This treatment is useless if the die has already cracked. In fact, the equipment often follows the procedures for placing speedy service with the consequent and severe thermomechanical stresses that undergoes during its life path. The material is thus subjected to a thermal stress proportional to the speed of heating, the amplitude of trips and thermal non-uniformities. At the end of sampling, the non-elimination of the internal tensions/stresses will be added to those of the subsequent reboots and made arrangements, so as to exceed the breaking load, causing the incidence of cracks that can be released quickly or be limited to multiple cracks as in the case of thermal fatigue.

After each batch, also, making operations such as washing of the die components, achievable with the use of equipment operating at high pressure with a spray or ultrasound, which allow to remove the residues of solidified metal, surface oxidation, lubricants, and performs an accurate control of the surface state. This is necessary to detect as early as possible any deterioration phenomena and to act appropriately with superficial and structural repairs of the same die.

4.4. Some considerations on the defects on castings

The defects are naturally produced in foundry processes for various reasons. The final properties and behaviour in operation of the castings depend on the microstructural characteristics and the presence of defects, which are the result of the different phases of the process, the physical-chemical properties of the alloy employed and the die and equipment configuration. Thus, extreme conditions that are created in the filling phase and the cooling process significantly affect the quality of the casting, so that the die can be considered as a process defects generator. Not only it will typically produce in average a percentage of 5-10% of the waste, but also the type, size, and the dangerousness of the defects are variable.

The surface defects of the die, caused by the various previously described damage mechanisms, are inevitably transposed to the castings.

The erosion wear is due to the impact of the flow of liquid metal in a turbulent wave and to the presence on it of non-metallic inclusions, removal of small amounts of material from the surface of the die, with the formation of pits (small cavities). The casting will present this situation: on its surface excesses of materials, of varying sizes, depending on the extent of wear will be present.

In addition, the presence of hard non-metallic inclusions, slag or granules coming from the gradual disintegration of the crucibles in Al, leads to one of the most common defects in diecastings, namely the hard points. One of the possible causes of the formation of hard points is associated to the gradual sedimentation of the heavy elements present in the starting ingot which feeds the melting furnace. In the time that elapses between the insertion of the ingot and its union, the elements such as Fe, Cr, Cu, Mn, Ni, contained in the liquid alloy in the crucible, lose their solubility with the Al, precipitating towards the bottom of the same, by virtue of their higher specific weight. Repeating the operation several times, it will go on to form a slime layer gradually increased the crucible bottom. The slime layer assumes a high danger if it is agitated (even with the insertion into the oven of a new charge), as it will be dispersed in the alloy, which will subsequently be withdrawn from the cup to feed a new working cycle with the fusion of a jet. Another possible cause of hard spots in the casting is due to the formation of oxides on the metallic walls of the die, following the reaction with the water of the lubricating release agent, and to their subsequent detachment in the casting, which will subsequently be withdrawn from the cup to feed a new working cycle with the fusion of a casting. Another possible cause of hard spots in the jet is due to the formation of oxides on the metallic walls of the die, following the reaction with the water of the lubricating release agent, and to their subsequent detachment in the casting, which will subsequently be withdrawn from the cup to feed a new working cycle with the fusion of a casting. Another possible cause of hard spots in the casting is due to the formation of oxides on the metallic walls of the die, following the reaction with the water of the lubricating release agent, and to their subsequent detachment in the casting.

The hard points, in a casting has to be subjected to subsequent machining in series and constitute a defect of considerable gravity, because they determine a rapid and irregular wear of the tools, and, in the most cases, their local fracture.

As a result of corrosion of the die, the cast will present a particularly rough surface, in a more evident manner in correspondence of the areas which are more corroded. In addition, it can also generate variations in size of the casting or local excesses of materials, corresponding to cavities or fissures that the corrosive phenomenon can generated in the die.

The metallization, tends to start easily in the areas of the die which are already damaged by erosion, corrosion and thermal fatigue, and in those regions exposed to the action of the liquid metal which is located at higher speed and temperature, door to a stratification of the alloy on the surface of the die, in the area gradually larger. Furthermore, these intermetallic phases may sometimes be detached partially and locally from the surface of the steel. These results in the casting are mostly related to the presence of roughness, smaller regions in relief and localized lack of material.

Extended use of the die leads to more or less deep formation of cracks on the surface, due to the repetition of stress-strain cycles, induced by quick changes in temperature as a result of alternating contact with the molten alloy and the release agent. Therefore, the liquid metal will penetrate into these cracks of the die giving rise, on the casting, tight surface projections distributed to form a mesh, which reproduces in negative.

The different damages lead to a loss of surface finishing properties of the casted product, which has certainly been subjected to additional mechanical finishing, with the unavoidable increase in costs.

The pieces or products which are usually under development in order to be appropriate, have to fulfil the requirements of the specific aim and have to be in line with the regulations and supply tolerances and whatever else agreed with the customer. As this task concerns, reference is made to the following legislation: "UNI 10429 - castings of cast aluminum alloys in pressure. General technical delivery"

To guarantee, during their production, that the pieces produced hold the required final characteristics, verifications, checks and tests has to be performed on them. These operations are generally referred to as quality control and are delivered by specialized personnel. To make sure that the production act is in conformity, it can be provided of verifying all the pieces in the control unit, or check one or more pieces after a certain number of components products as a statistical control. In practice, in the foundries the use of the second option is preferred, in order to prevent discontinues in the production route.

The most important controls and verifications which generally are performed can be listed as follows:

- control of the visual appearance of the pieces;
- structural integrity checks;
- mechanical-dimensional controls.

In cases where it is considered necessary, pressure tightness tests, tests of mechanical strength, machining tests and surface finish tests can also be performed.

The visual inspection of the pieces is carried out typically by the operator assigned to the machine, and purpose of this examination is the visually ascertain of the quality of the pieces produced and in particular:

- fullness in every detail, the absence of missing parts;
- the surfaces of the presentation state;
- the absence of visible defect, or its extent and severity in the case if one observes their presence.

The structural integrity checks are important, because they depend on properties such as mechanical strength, pressure sealing, the status and functionality of the surfaces subjected to mechanical machining. The absence of porosity, of cavities, of blowholes, of structural discontinuities will be checked. If present, one has to be sure of their entity, in order to determine their acceptance or the rejection of the produced pieces. The structural integrity check may be carried out by means of:

- destructive examinations: machining, fracture, sectioning;
- non-destructive testing: X-ray fluoroscopy or radiography.

The mechanical-dimensional control of the parts has the goal to verify that the shape and any size (geometric dimensions, thicknesses, diameters, depths, distances of the holes, etc.) are those provided by the design or within the tolerance agreed with the costumer. The means to achieve the purpose are numerous and range from the most basic one to the most sophisticated ones:

- measuring instruments in common use: lines, gauges, micrometers, comparators;

- measuring instruments technologically advanced: electronic gauges and lines of programmable computerized measuring machines;
- control of all the tools which are specifically built for verification for a specified component.

Among the characteristics required to diecast components, depending on their specific application, there is the seal. It has the ability to contain a fluid leak-free, and is required for certain pieces that should fit for their specific function. The seal in a component is directly proportional to the structural integrity and health of the pieces itself and of its thickness. When the piece requires the pressure seal, this gas to be ascertained with sealing tests, e.g., air-water: this test consists generically in filling of compressed air the piece, which it is immersed in water and the seal is verified if it does not produce bubbles in the water. Alternatively, one can use technologically more sophisticated systems, in which the seal is verified thanks to electronic circuits, so as to eliminate the discomfort of the operation carried out in water. In special cases, the losses due to micro-porosity can be eliminated by the impregnation of the piece (with special epoxy resins).

4.5. The technological degradation of the dies

The limitation of the life of the die is due to several factors, since its surfaces are subjected to high temperatures when they are in contact with the liquid metal, and between a molding cycle and the other the die undergoes a drastic cooling. The alloy injected, moreover, is subjected to compaction pressures (35-100 MPa), and then the die surfaces are subjected, in addition to thermal shocks, even in continuous alternation of compression and release. The thermal shock and mechanical stress are among the main causes that limit the life of a die. Other limiting factors are erosion, corrosion and metallization. In fact, the alloy is injected at high speed into the die cavity and the flow of metal causes abrasions, which, together with the cavitation phenomena, removes part steel die, altering the shape and dimensions, or by creating undercuts that counteract the removal of the piece. The extent of erosion and corrosion is relative to the type of alloy used, its temperature and the scroll speed.

The geometry of the die cavity has a great influence on the life of the same. In correspondence of the edges, holes, ribs or studs, generation of heat concentrations occurs, that induce locally from tensions in the construction of the die material. In these areas the temperature of contact between the injected alloy and the steel of the die also increases. The combined presence of tensions and increase in the contact temperature causes the thermal fatigue cracks to occur first at these points than on the flat surfaces. The concentration of thermal fatigue cracks at corners increases the risk of complete split of the matrix block.

The service life of a die is then conditioned by:

- factors related to the geometry of the die;
- factors due to the diecasting process.

When corrosion occurs, this is a result of the dissolution of the constituent material of the die in liquid Al, with the consequent formation of intermetallic phases. The corrosive wear result is the formation of craters and small protuberances of Al-Fe intermetallic compounds on the die wall. Such damage can then result in the metallization and the loss of the die surface quality.

The corrosion of steels in contact with the die and with the liquid Al is the result of the dissolution and diffusion of Fe in the molten alloy, and the driving force of these phenomena is the difference of chemical potentials of Fe and Al between the matrix and the liquid bath. One limitation to the problem would be to increase the content of Fe in the alloy so as to decrease the difference of chemical potentials, but the Fe affects the formation of the crystalline alloy with bigger grain size, casing the fragility of the casting and favoring the formation of intermetallic phases. For all these reasons, the limit for Fe, in percentages, is about 1%. The phenomenon of corrosion depends mainly on the temperature, which plays a fundamental role on the solubility of the various chemical elements. All of the corrosion reactions, in fact, are matter of thermodynamics, miscibility and the formation of various compounds ($FeAl_2$, Fe_2Al_5, $FeAl_3$) in accordance with the equilibrium diagram of Al-Fe, which has a fundamental role on the solubility of the various chemical elements.

The erosion is the damage caused by the impact of the molten alloy on the die surface, and the wear conditions become progressively more severe with increasing temperature and speed of the metal. The erosion is mainly due to the inclination of the liquid Al flow (it is maximum when it is perpendicular) and the eventual presence of solid particles of high hardness, such as non-metallic inclusions. In addition, when there are variations and inequality in the section of the casting channel, the air and the gases dissolved in the fluid form bubbles, which are then reabsorbed. In the presence of these gaseous bubbles, generally spherical, cavitation phenomena occur on the surface of the die: the implosion of the same freedom to a great energy that brings the fluid stream to flow non-uniformly, but with a series of pulsations and vibrations that generate waves pressure and depression with consequent erosion of the die. The erosion is due to the removal of small quantities of material and, therefore, the result of the action erosive results in the formation of small cavities (pits) or in relatively large portions of eroded surfaces.

The metallization (soldering or die) is the result of interaction between the liquid Al and the steel made up: the alloy, in fact, joins on the surface of the die during the die casting process, and remains adherent even after the extraction of the casting. Such damage leads to an inevitable reduction in the surface finishing properties.

The liquid metal is injected, through the sprues, inside the die at high speeds, pressures and temperature, and considering that each molding cycle typically lasts less than one minute, the surface of the die, is subjected to repeated blows by the alloy and will inevitably be subjected to excessive wear. Al attacks the weakest regions, , steel, and thus forms the dimples. The molten Fe coming from steel diffuses in Al leading to the formation of intermediate layers of binary phases Fe-Al and ternary Fe-Al-Si. The region covered by the presence of intermetallic widens progressively.

Mechanism of metallization: it is not of electrochemical type, but purely based on diffusion and chemical reactions among the constituent elements of the die and the liquid metal.

In the first stage, the molten Al comes into continuous contact, at each molding cycle, with the die surface. The molds for steels are generally heat-treated and subjected to a double (or triple) tempering to a hardness of approximately 48 HRC. During the casting process, the molten aluminum attack the weakest regions of the surface of the matrix, i.e., the regions between the slats of martensite and the particles of carbides. When the alloy erodes such areas, there is the formation of a primary solid solution of iron, represented by α-Fe and of hemispherical shape dimples, which progressively expand laterally and in depth, with grains release superficial. Subsequently, the aluminum reacts with the surface grains loose from the matrix, and, to the surface of the pits, are formed binary iron-aluminum phases, such as FeAl, FeAl2, Fe2Al5, FeAl3. The formation of these successive layers of binary compounds is due to the reaction with the Al that is continually refilled and the diffusion of the iron outside of the steel surface. In the next stage, the FeAl3 phase reacts with the Al and Si of the liquid metal to form a ternary phase α- (Al, Fe, Si). The intermetallic layer that is formed in this stage typically has a pyramidal morphology. In the next stage, the FeAl3 phase reacts with the aluminum and silicon of the molten metal to form a ternary phase α- (Al, Fe, Si).

When the molten Al volume is ample, the reaction between the intermetallic phases and the liquid dominates the diffusion of the Fe from the surface of the die. The ternary phase has greater thickness with respect to the layers of the other phases, although the total thickness of the intermetallic layer is governed, however, by the spread of Fe.

The Si and other minority elements (Cr, Mn, V) coming from the die and the Al alloy precipitate at the grain boundary phase of the Fe2Al5. It can be found, in addition, are large precipitates and rich in Si at the interface between the binary phases and ternary ones.

As soon as the intermetallic layer is formed on the die surface, the Al excess is stratified, consolidating and remaining adherent even after the expulsion of the jet, thus leading to a lower surface finish of the pieces produced, as well as to possible problems in the extraction.

The bonding is mainly due to the stop of the reaction taking place between steel and Al, and the effect of the protruding surface energy intermetallic layer in the liquid metal. Another possible explanation of the stratification of the intermetallic alloy on the protuberances is attributable to their low coefficient of thermal conductivity than that of steel.

Over time, the dimples widen and unite with each other, and, once formed the intermetallic layer, they begin to extend only parallel to the surface of the die, so that, when it is replenished with new molten Al, it shall enter into contact with the steel only in correspondence with limited gaps and crevices between adjacent pits. The welded Al re-melting when comes into contact with the new volume of the liquid alloy is not so possible, since the contact times are very short.

4.5.1. Influence of process parameters

The main factors that influence the reactions at the interface of the Al-die molten steel are: chemical composition of the alloy and steel, alloy temperature, die temperature, operating parameters (injection pressure and speed).

At the level of chemical composition of the alloy, Fe has a great effect on the metallization: increasing the content at the saturation limit (which falls in the range of 0.9-1.15% by weight), it reduces the metallization. In alloys which instead have a low Fe content (around 0.4% by weight), one can use Mn (0.85 wt%) in higher percentages to control the metallization and thus can reduce the content of Si (around 7% by weight) to enhance the chemical activity of Mn and Fe in the alloy, and then reducing the effect of the damage.

With regard to the composition of steel for die, it can be modified to better withstand to a liquid Al attacks.

The temperature of the alloy is a critical factor in the formation of hot spots on the die surface. For this, one has to maintain the melt temperature as low as possible before casting, compatibly with other process conditions in order to minimize the presence of harmful hot spots for the die surface, which will be precisely the most stressed one and it will show a higher tendency to soldering in such areas.

The mold temperature is another important variable that can promote damage: higher temperatures increase the activity of the surface atoms, increase the coefficients of diffusion and consequently increases the reaction rate, and everything will be reflected at an early and rapid metallization. In addition, high temperatures can have a tempering effect on the steel surface, promoting wash-out phenomena and soldering as a result of softening.

As concerns the effect of the pressure of injection, it is necessary to consider both the mechanical action and chemistry action. In fact, on the one hand, any coatings or layers present on the oxidized surface of the die can be partially removed by the violent molten Al attack injected at high pressures, and then the alloy can come into direct contact with the steel. Furthermore, the energy of the alloy, if injected at high pressures, also increases the number of atoms in a state of activation. Thus, the metallization is carried out much more easily in the presence of high injection pressures. Similarly, high injection speeds accelerate the processes of erosion and consequently lead to metallization.

Finally, also the state of the surface is a critical variable in the prevention of the aforesaid damage. On macro-scale, the areas of the die, which protrude into the molten alloy, are much more likely to metallization because of the increased temperature. The same phenomenon will also occur on micro-scale: that is, on a rough surface, the affected small protrusions of a temperature higher than any localized area on a smooth surface. They are therefore preferable smooth surfaces, but obviously not below the limit required to ensure that the lubricant adheres to the die and to satisfy its beneficial role. An optimal condition of the surface can be obtained by means of surface modifications or coatings that harden the steel surface and slow down the erosion.

4.5.2. Thermal Fatigue

With thermal fatigue it is defined the process of damage that can be found in the die which are subjected to thermal cycles, or to cyclic heating and cooling, according to thermal waves of limited penetration.

Generally, steels for dies retain a coefficient of thermal conductivity that increases with temperature: if the coefficient of thermal conductivity is too low, the heat exchange is slow, so the die cannot be cooled intensively, otherwise it would increase too much the difference in temperature between surface and core, with excessive risk of fracture due to thermal shock. The thermal shock resistance is proportional to the toughness of the steel, which grows rapidly with temperature: fractures in fact are more frequent in harder steels and less tenacious, that is, in those where the propagation of cracks under tension is easier. In order to reduce the thermal shock, it is necessary that the heating and cooling periods to be as short as possible.

The thermal cycles are able to generate on the surface of the alternation of compression and traction tensions: the rapid heating of the surface of the die during the injection of the alloy generates compressive stresses in a thin surface layer and traction in the sub-surface layer, due to the thermal expansion of the heated part, prevented by the mass of the still cold die. These tensions are released as soon as the limit is exceeded yield hot compression to the surface or that hot traction to the underlying fascia, through micro-plastic deformation of the steel. These mechanical stresses that are generated as a result of thermal shock in the long run lead to the formation of thermal fatigue cracks.

The cracks tend to expand over time and inevitably lead to the destruction of the die. Penetrate generally in a direction orthogonal to the surface, and then turn to the contour of the grains, which are often undermined with strong local enlargement of the crack.

Furthermore, this damage will affect inevitably the surface of healthcare casted products. The thermal fatigue phenomenon is favoured by the corrosive liquid Al with which the die comes in contact, and the concentration of efforts due to accidental notches, grooves such as mechanical processing, to data notches from the geometry of the piece. Therefore, it is useful to smooth the surfaces to eliminate any roughness or deep channel, and to avoid as much as possible rapid changes of section.

The trigger of the cracks is also favoured by the layer of oxide which is formed on the metal surface of the die as a result of thermal cycling. This oxide layer has low thermal conductivity, higher volume and is particularly fragile. Tensile stresses that develop in the cooling phase determine the local formation of cracks in the oxidized layer, which, due to its intrinsic fragility and of the thermal expansion difference with respect to the substrate, leads to the opening of the same. These cracks act as preferential channels for the introduction of liquid alloy and oxygen, which at high operating temperatures cause oxidation of the cavity of the crack itself.

4.5.3. Thermal stress on the surface

During casting, the die is subjected to thermal gradients, since the core remains at a lower temperature compared to the surface. The heat flows from the alloy to the die, heating the surface, when they are in contact, while the entire die cools during the removal of part of the heat in the process of extraction of the casting and lubrication. The thermal gradients lead to dimensional changes that generate stress and deformation. There are several stress-strain cycles that occur during casting.

The nucleation and growth of cracks by thermal fatigue is well described by the Coffin-Manson and Solomon equations. These models indicate that the number of cycles that leads to the nucleation of cracks, as well as to their growth, varies exponentially with the magnitude of the plastic deformation.

5. Hot forging and some recommendations during its application

In hot forging process, a portion of the starting material, for example a billet, is plastically deformed by means of two tools, so as to obtain the chosen configuration. In this way, a piece with a simple geometry is transformed into a component with a complex shape, to which the die impresses the desired geometry and imparts pressure to the material through the surface of the instrument. As a result, the forging process offers potential savings in energy and material, especially in the case of medium and large quantities production, allowing the cost of tools to be amortized earlier. The physical phenomena that characterize the forging operation are difficult to express in a qualitative way. The flow of the metal, the friction at the material-die interface, the heat generation and its transfer during the plastic flow and the relationships between microstructure, properties and process conditions are difficult to predict and evaluate.

Usually, in the forging process several steps (pre-forming) are necessary to transform the simple initial geometry into the final complex one. For a given operation (pre-forming or final forging) certain guidelines has to be followed:

> a) establish the kinetic relationships (shape, velocity, strain rate, deformations) between deformed and non-deformed parts, or predict and study the flow of material;

> b) establish the limits of formability and productivity, namely to determine if it is possible to form the part without causing internal or superficial cracks;

c) predict the forces and stresses necessary to perform the forging operation and as a consequence to be able to design and select the necessary equipment and dies.

The development process refers to precision forging, which seems to be able to remedy the problems from economic and environmental points of view. It is possible to state, in an indicative way, that about half of the cost of a forged component is attributable to the acquisition of the starting material. The use of a precision forging allows a saving of about 15% material, so the cost of the final component can also be reduced by about 7.5%. Usually, the precision forging is achieved through the use of a closed die in order to guarantee a constant quality of the work and to avoid possible damage to the forged piece, following cutting or drilling. The main target of enclosed hot forging technology (hot forging with closed die) is to improve the utilization of the material in an optimal process and lowering the energy necessary for deformation. This aim can be achieved by performing the necessary deformation in the single directions (thus saving energy) and automating the process.

5.1. Plastic deformation and forging

The key to the success of the forging process, e.g., obtaining the desired shape and properties, is the full understanding and control of the flow of the material. The direction of metal flow, the level of deformation and the working temperatures play a fundamental role in the final properties of the forged component. The flow of the metal determines both the mechanical properties referred to the local deformation, and the formation of defects, as cracks or folds, on the surface or below it. The local flow is, in turn, influenced by the multiple process variables. To study the plastic deformation of a metal it is necessary to consider homogeneous or uniform deformation conditions. The stress of yielding a metal under uniaxial conditions, as a function of deformation, velocity of deformation and temperature, can be assumed as flow stress. The material begins to deform or flow plastically when the stress applied, under uniaxial stress or compression, reaches the yield value or flow stress. Flow stress is very important, since in the metal forming processes the loads and stress depend on the:

- geometry of the piece;
- friction;
- flow stress of the material to be deformed.

It is possible to state that the factors that influence the flow stress of a metal can be divided into two groups:

- variables separated from the plastic deformation process, such as chemical composition, metallurgical structure, phases, grain size, precipitates and previous deformations;
- variables related to the plastic deformation process, such as process temperature, amount of deformation and deformation speed.

In this way, the flow stress σ_m can be expressed as a function of temperature, θ, deformation, ε, velocity of deformation, $\dot{\varepsilon}$, and microstructure, S. Therefore, for a given microstructure, a given heat treatment and preventive deformation can be described simply by the function: $\sigma_m = f(\theta, \varepsilon, \dot{\varepsilon})$.

In hot forging processes above the recrystallization temperature, the effect of deformation on the flow stress is insignificant and the influence of the deformation speed becomes enormously important. The degree of dependence of the flow stress from the temperature varies considerably depending on the material. Thus, temperature variations during the forming process can have different effects on applied loads and metal flow for different materials.

As an example, one can consider, a temperature drops of 100 °F (55 °C), a route from 1700 to 1600 °F (from 925 to 870 °C) in the working range of an AISI 4340 steel leads to an increase of about 15% of the flow stress to be applied.

The compression test can be used to determine the data concerning the flow stress, relationship between true-strain and true-stress, for many metals at various temperatures and strain rates. In such tests, the flat plates and/or the cylindrical sample are kept at the same temperature so as to prevent cooling of the die and the consequences that this would have on the flow of the metal. In order to be applicable without corrections or errors, the cylindrical sample has to be pressed in such a way to maintain a constant stress state on the whole pieces. For this step the use of a suitable lubrication, glass powder for high temperature steels, can be helpful. The load and strain, or height of the sample, are measured during the test. Starting from this information it is possible to calculate the flow stress at each instant of the deformation or by increasing deformation. In the case of the tensile test the following relationships, related to the deformed part, are valid in the case of uniform compression:

$$\varepsilon_m = \ln(h_o/h) = \ln(A/A_o)$$

$$\sigma_m = L/A$$

$$A = A_o(e)^{\varepsilon_m}$$

$$\dot{\varepsilon}_m = d\varepsilon/dt = (dh/hdt) = V/h$$

where V is the instantaneous deformation speed; h and h_o are the initial and instantaneous heights, respectively, and A and A_o are the initial and instantaneous surface areas, respectively.

While the values of flow stress obtained by tensile tests with high deformations require a correction due to the phenomenon of necking, the compression test, which can be carried out without the occurrence of the above mentioned phenomena up to a reduction in height of about 50%, is mainly used to obtain data concerning the flow stress during metal forming operations. At room temperature, the flow stress of many metals, excluding Pb, is only slightly dependent on the velocity of deformation. Therefore, some machines or presses can be used for the compression test. Proper lubrication of the plates is usually performed (i) using lubricants such as teflon, molybdenum disulfide, or high viscosity oils and (ii) using samples with spiral surface hollows, on the upper and lower faces, containing lubricant. During the process of forming at high temperatures (T> $T_{recrystallization}$) the flow stress of any metals is extensively dependent from the velocity of the deformation.

Therefore, when possible, the hot compression tests are carried out by means of instruments capable of providing a velocity-displacement profile such that the condition $\dot{\varepsilon}_m$ = (velocity/ height of the sample) is maintained for the duration of the test. To achieve this aim, mechanical presses or hydraulic machines that are programmable for testing are used.

The test is carried out in a so as to maintain uniform and isothermal the compression conditions. The sample is lubricated with appropriate lubricants (for example, graphite oil for temperatures above 400 °C and glass for temperatures above 1260 °C). The equipment and the samples are heated up to the test temperature before it takes place. The higher σ_m refers to more resistant materials. At the temperatures at which hot working is carried out σ_m increases with increasing $\dot{\varepsilon}_m$ and with decreasing temperature, θ. In the case of constant $\dot{\varepsilon}_m$, the curve

σ_m vs. ε_m firstly grows, then it decreases as a result of the internal heat generation and thermal softening. Usually, during the analysis the test temperature is not constant in the strict sense. In fact, due to the plastic deformation, a temperature increase takes place, $\Delta\theta$, which can be estimated:

$$\Delta\theta = (A\varepsilon_m\sigma_m)/(c.\rho)$$

where A is a conversion factor, c is the thermal capacity and ρ the density.

$$\sigma_m = c(a+\varepsilon_m)^n$$

In the case of materials sensitive to deformation speed, the most used expression is:

$$\sigma_m = C(\dot{\varepsilon}_m)^m$$

The coefficients C and m will be obtained at various temperatures and deformations, so that they will have different values at a given temperature, depending on the deformation. To found forces and stress in practical forming operations, it is often sufficient to specify the mean or maximum value of σ_m to be used in the equations to obtain the maximum forming load. In some practical cases the use of a constant average value for σ_m is justified.

5.2. Stages of the hot forging process

Hot forging is defined as the process conducted above the recrystallization temperature so that the metal recrystallizes during the deformation (often however the strain rates $\dot{\varepsilon}$ are so high as to make the recrystallization difficult, so the piece is located in the annealed state). This process exploits the decrease of the high temperature flow stress to reduce:

- the forces to be imprinted on the die to have plastic deformation;
- the size of the equipment;
- the requested power.

By hot forging mechanical parts or machines that require high mechanical strength (crankshafts, disc tools, gears, pistons, etc.) are produced. The forging process is carried out at a temperature, which is a function of the melting temperature of the material, T_m, which can be approximately obtained as: $T/T_m > 0.6$, while the process takes place in the following phases:

1. cutting the billet;
2. heating of the material;
3. molding operation;
4. subsequent processing (deburring and tool machining);
5. thermal treatments (if any);
6. sandblasting.

5.2.1. Cutting the billet

The standard amount of material to be forged is obtained by cutting semi-finished steel products such as billets, usually obtained by rolling, or bars with a square section, so as to obtain exact volume and height. The cutting operation usually produces a uniform cutting edge with little or no damage to the microstructure in the immediate vicinity. Cutting the billet using cutting tools is a process without waste of material and with much higher productivity compared to other methods (flame cutting, abrasive cutting). Billets and sections of bars are sheared between the upper and lower blades of a machine, which has only the upper movable blade. Other cutting methods, such as impact cutting, use a tool similar to a knife, moved horizontally, to shear the bar portions. In this case, the blade plastically deforms the material until the deformation limit is reached in the cutting zone; at this point cracks appear due to the cutting operation and there is separation. This operation, however, gives rise to a damaged or deformed surface. Many materials cannot simply be cut into portions of exact height and volume. Many times, especially in the case of high-strength steels with tensile strength higher than 400 MPa, a preventive heating between 300 and 400 °C is necessary, in order to eliminate the risk of cracking. Starting from this phase the forging process is strictly automated, so as to guarantee constant quality.

5.2.2. Heating of the material

Starting from this phase the forging process is strictly automated, so as to guarantee constant quality. A pieces is conducted, manually, to the entrance of the oven, where special loading robots transport the billets inside the induction furnace. The hot molding process requires heating at high temperature and uniformly of the billet in the two direction (longitudinally and transversely). Generally, the starting temperature is room temperature and it is necessary to heat up above the recrystallization temperature of the material. In the case of steels, a typical thermal interval in which it is correct to operate is between 1100 and 1280 °C, depending on the composition. There are various ways to heat a steel billet before hot pressing (induction heating, infrared heating, electrical resistance, gas or gas furnaces), but the one with the greatest advantages is induction heating. First of all, the induction heating systems allow to quickly creating high heat intensity in a given area of the material. Moreover, such systems have a rapid start-up phase, since they do not need a pre-heating time, as is the case for fuel ovens. Other advantages, briefly, are: the very short heating time, compared to other methods, the simplicity and control and reproducibility of the process. This last feature highlights the possibility of automating the process. It is also an efficient method from the energy and environmental point of view, because there are no unsafe emissions. Also its impact on the physical characteristics of the piece is very limited: poor polluting and minimal decarburization of the billet compared to other methods. This heating method results in the lower formation of surface oxide. At the exit point of the furnace, special extractor robots move the luminescent pieces moving them towards the forming area, where the molding takes place. This operation has to be performed quickly in order to avoid a rapid lowering of the temperature and a consequent bad forging.

5.2.3. Molding operation

The molding operation is the central phase of the hot forging process. A robot moves the pieces coming out of the induction furnace and loads them on a special machine called a press (hydraulic, screw). The presses are classified according to the: closing force and closing energy.

For a given geometry, the load value varies with the flow stress of the material and the friction conditions. During the forming operation the machine must produce the maximum load, and therefore the energy required by the process.

Depending on the deformation speed and the thermal effects, different loads applied for different machines are required for the same forging process. Before molding takes place, it is necessary to preheat the dies, which during forming will reach temperatures between 400-500 °C. This procedure is necessary to eliminate the risk of cracking due to thermal shock. The robot then places the luminescent billet on the lower die. Once the molding has been carried out, the molded piece is removed from the press by automatic robots or by an operator using a special tool.

5.2.4. Subsequent processing (deburring, tooling)

Once the geometry of the molding piece has been obtained, is completed by means of auxiliary operations such as deburring and tool machining.

The deburring process consists of the finishing and sanding of the piece, as well as the elimination of the metal burrs that form on the edges of the component during molding. Generally suitable deburring presses are used. Tool machining includes cutting, milling and turning operations.

5.2.5. Thermal treatments (if any)

If necessary, heat treatments are carried out to improve the mechanical characteristics of the pieces and to eliminate any residual stress, if present. The most used thermal treatments are annealing and hardening. It is also possible to perform thermochemical cementation or nitriding treatments to obtain suitable surface characteristics.

5.2.6. Blasting

The sandblasting technology makes possible to eliminate any deposits and oxide layers present from the surface of the pieces. Special machines are used which use a jet of sand and air, sent to the piece at high speed. The success of the hot forging operation depends on a series of variables, and it is possible to make a logical subdivision considering the variables that get involved respectively before, during and after in the forging process.

Once the molding has been carried out, the molded piece is removed from the press by automatic robots or by an operator using a special tool.

The main variables that occur before the forging process are:

- work piece: it is necessary to choose the appropriate geometry and material and to know the characteristics of forging (flow stress) and all the earlier treatments and processes which the material have been submitted. It is also necessary to calculate precisely the volume of the billet to be forged in order to avoid defects or irregularities in the forged piece;

- mold: two important factors have to be considered, namely the production method and the design. Different mold production methods provide different dimensional accuracy and surface roughness. The design of the mold affects the method used, depending on the complexity and the geometry to be built;

- process variables: this is the set-up and design of the mold and the number of steps of the forging process. By mold set-up one means the correct positioning of the equipment by molding and periodic corrections of the right positions during forging. The design and the dimensions of the mold has to be greater than the path to be performed so as to be able to control the elastic deformation of the mold itself during the pressing;

- machinery: the accuracy of the process depends on the accuracy of the equipment used. This in turn is a function of rigidity, maneuverability and distribution of the load and of the energy.

The main variables that occur during the forging process:

- workpiece temperature: this is the most important variable. The temperature must be monitored carefully from the billet outlet from the induction furnace. During the first forging steps there is a decrease in temperature due to contact with the molds. During forging, the working temperature influences the flow stress and therefore the required load and energy. Low temperatures, in fact, cause elastic deformations of the molds, to which an excessive load is applied;

- oxidation and decarburization: the phenomena of decarburization and oxidation can derive from different conditions. The main elements are: temperature, temperature maintenance, working atmosphere and material. Steel starts to oxidize from 200 °C, but severe oxidation, with loss of material, starts at 800 °C. To minimize oxidation it is necessary to quickly heat the billet up to the forging temperature, without maintaining it for a long time at high temperature. Induction ovens are suitable for this purpose. To avoid both oxidation and decarburization, the protective atmosphere is an excellent solution. In extreme cases the decarburized layer can be removed by tool machining;

- billet volume: variations in the volume can cause variations in the load and energy required for forging and changes in the final dimensions of the product (especially the thickness). The billet cutting machines allow to obtain the volume necessary for forging, monitoring the bar diameter and cutting length;

- lubrication: lubrication during forging is important, as it reduces frictional wear at the part-mold interface, allowing it to reduce the load and energy required for forming. The lubrication also allows to reduce the temperature of the mold and improves the extraction of the forged piece even with small angles of the

produced bare. These advantages are obtained only in the case in which there is a suitable lubricant layer, uniformly distributed and resistant to high temperature;

• mold wear: the mold is the fundamental element of the process, as it affects the precision of the forging as well as the cost-effectiveness of the single piece. 70% of the molds are replaced due to wear problems, because the tolerances are no longer respected. Therefore, reducing the phenomenon of frictional wear, the life time of the molds is prolonged and the efficiency of the process is improved. It has been estimated that overheating of the molds causes a reduction in the life time of up to a third. For this reason, a lowering of the forging temperatures avoids this phenomenon, as well as un-favoring the formation of oxide on the surface of the piece, which would contribute to the increase of abrasive wear.

Variables that occurs after the forging process

The main variables that occur after the forging process are:

• deburring: the deburring process allows to eliminate the metal residues that surround the forged piece in order to improve dimensional accuracy. In some cases, it can give rise to distortions or put at risk the dimensional accuracy of the pieces produced;

• heat treatment and handling: sometimes the heat treatments, usually performed to obtain an optimal compromise between strength and toughness, are incompatible with the tolerances required for precision forging or can cause decarburization, oxidation or distortion. Batch treatments have also been avoided. The solution to these drawbacks can be the use of protective atmospheres;

6. Some considerations on X-Ray diffraction and its use for metals and alloys - Examples of real case studies

6.1. Generalities

X-ray diffraction is a non-destructive characterization method used to obtain structural data on an atomic scale. Through this technique is possible to analyze crystalline materials and amorphous materials too. The most important purposes when one uses this technique are to define the structure of "natural" and of "manufactured" samples, in order to determine some characteristics of the material, like: chemical and phase composition, crystal orientation, crystallite size, lattice parameters, surface and interface roughness, etc.

In the case of metals and metallic alloys the most important features which can be determined analysing the peak profile of the X-ray spectrum are related to the determination of the crystallite size, the solid solution inhomogeneity, temperature factors, micro-strain, non-uniform lattice distortions, defects, dislocations, grain surface relaxation.

Metals and alloys are made up by atoms, which are held together by strong, uniform and delocalized bonds, which give high mechanical resistance, allowing the movement of atoms. The last issue is directly correlated to their ductile property and on why they can be deformed without fracture and can be exploited in case of the production of sheets, wires, etc.

The electrons are attracted to many atoms, they have considerable mobility determining the good heat and electrical conductivity. In a liquid metals the atoms are randomly arranged and they are relatively free to move. During cooling the metals reorganize to form an ordered crystalline structure.

In metals and alloys one can have a non-uniform lattice distortion because of the presence of inclusions, oxides, surface tension, different morphology and shape of the crystals made up the structure, interstitial impurities. X-ray diffraction peaks are formed by constructive interference of a monochromatic beam of X-rays scattered at specific angles from each set of lattice planes in a material. The peak intensities are determined by the atomic positions within the lattice planes. The distance (d) between the atomic layers of the crystals are variable (different planes have different spacing) in the X-ray spectrum, any intensities produced correspond to a precise compound.

A substitutional impurity atom is an atom of a different type than the bulk atoms, which has replaced one of the bulk atoms in the lattice. Substitutional impurity atoms are usually close in size (within approximately 15%) to the bulk atom, e.g. Zn atoms in brass. In brass, Zn (radius = 0.133 nm) replaces some of the Cu (radius = 0.128 nm). Vacancies are unfilled places where an atom should be, but is absent. Common feature particularly at high temperatures when atoms are frequently and randomly change their positions leaving behind empty lattice sites.

Interstitial impurity atoms are much smaller than the atoms within the matrix. They fit into the open space between the bulk atoms of the lattice structure, e.g., in steel the C atoms (radius= 0.071 nm) are added to Fe (radius= (0.124 nm).

The degree of the strength imparted by the alloying element depends on the relative difference in size between the solute and solvent. The crystallite size is different than particle size. A particle can be made up of several different crystallites and the crystallite size frequently matches grain size, but there are exceptions.

In case of a comparison with standard reference patterns and measurements, the experimentally obtained pattern allows identification of the material and lead to define cell parameters and space group even in case of unknown material. X-ray methods for crystallite size determination are applicable to crystallites in the range of 2-100 nm. The diffraction peaks are very broad for crystallites below 2-3 nm, while for particles with size above 100 nm the peak broadening is too small. If the analyzed crystals are free from micro-strains and defects, peak broadening depends only on the crystallite size and on the diffractometer characteristics.

The three-dimensional structure of non-amorphous materials is defined by regular, repeating planes of atoms which form a crystal lattice. When a focused X-ray beam interacts with these planes of atoms, part of the beam is transmitted, part is absorbed by the sample, part is refracted and scattered, and part is diffracted. When an X-ray beam hits a sample and is diffracted, we can measure the distances between the planes of the atoms that constitute the sample by applying Bragg's Law:

$$n\lambda = 2d \sin\theta$$

where:

n is the order of the diffracted beam; λ is the wavelength of the incident X-ray beam; d is the distance between adjacent planes of atoms (the d-spacings); θ is the angle of incidence of the X-ray beam.

In case one knows λ and measure θ the d-spacing can be calculated. The characteristic set of d-spacing and their intensity generated in a typical X-ray scan provides a unique fingerprint of the phases present in the sample.

Before X-ray diffraction measurement one have to prepare the sample to be analyzed. Usually, for metals and alloys bulk material is used for the analysis, but in case of sintering, it would be interesting to evaluate also materials in form of powders. Bulks material requires smooth surface, after polishing the sample should be thermal annealed to eliminate any inconvenient, e.g., stress, induced during polishing. During the data interpretation it is important the peak position, its width and its intensity. The effect of lattice strain on the diffraction peak position and width is clearly visible on the spectrum: a sharp and intense peak indicate a material with no strain, a shift to lower angles, when d-spacing is uniform, is sign of the presence of a uniform strain in the material, while the peak enlarges and the d-spacing is uniform the material is non-uniformly stressed. The Full Width at Half Maximum (FWHM) is important for investigations about the particle or grain size and on the measurement of the residual strain, significant features in metallurgy. The measurement is very sensitive to the variation in microstructure and stress–strain accumulation in the alloy. Through FWHM measurement information about the surface state of alloys is obtained and this can be associated to the grain distortion, dislocation density and residual stresses within the metal or alloy.

Actually, there are at least two main methods for investigation of metals and of metallic alloys.

1.-Metallography, which is usually carried out through the observation of polished and etched surfaces;

2.-Analysis of the cooling curves, method for observing the discontinuities that can indicate some kind of phase changes.

Both of these methods require considerable skill and experience and the results were not always unequivocal.

Introduction of X-ray diffraction techniques provides a much clearer, simpler and more objective way of investigation and leads understanding the atomic arrangements of the materials involved. Therefore, the different characterization methods which are usually used for the material description are complementary and their use gives a much clearer image about the performance of the material.

In metallurgy, X-ray diffraction is very helpful for the analysis of standard metal alloys and comparison of manufactured alloys to the standard, to monitor the phase transformations during different thermal treatment (is not practical for pure metals, where there are no any phase transformation), to investigate alloys prepared in special conditions, e.g. in case of surface modification (nitriding, coating, chemically altered surface, etc.), to analyze the impurities, e.g. Fe in Al-based alloys, where is dangerous, to identify the presence of intermetallic compounds in Al or Mg-based alloys, to measure the Fe and Mn ratio, helpful in Al alloys foundry, to determine the repeatability statistics of metallic alloy, to measure the residual stress in alloys, important features when the alloy was deformed, thermal treated, etc.

6.2. Real case studies

To fix the theoretical ideas, in the following, some real case study is presented.

The first one is related to the analysis of standard metal alloys. By comparing the spectra obtained for different alloy composition to be identified the purity of the alloys compared to the standard one and chooses the right one for example for a specific determined application.

Other example is concerning the detection of different phase, for example, CuO using a comparison method and a standards and one can detect the phase evolution during annealing for different times leading to the selection of the most promising superconducting thin films.

Other interesting case is about some alloys, with the same composition, 70Cu-30Zn- brass, where one of them is cold-rolled and the others are annealed at different temperature. The as-rolled alloy shows a broaden peak and as the annealing temperature increases the crystallinity of the alloy increases as well. At the highest annealing temperature instead of the one large peak one can observe two sharp and well defined peaks, sign of a crystalline structure. The modification of the peak shape is sign of a variation of the grain size within the alloy. Increasing the annealing temperature, the intensity of the peaks gradually increased and the FWHM of the peaks gradually decreased.

The study of the development of intermetallic phase is other task on the route of the characterization of the alloys. Firstly, the alloys are investigated by scanning electron microscopy, observing their morphological appearance. The growth of intermetallic phases as a function of the composition and more in detail when Mg content increases, is different: as Mg content increases the secondary particles content increases, since the possibility to have a higher probability to have a chemical reaction and the formation of the new grains is higher, as reported in Figure 1. When, one observes the microstructures, some previous knowledge is very welcome.

Following X-ray diffraction, the X-ray spectra reported in Figure 2 is obtained. Each compound has a precis 2-theta point and a relative intensity.

The analysis of the formation and development of the new grains, in particular related to the growth of Mg_2Si and $MgZn_2$ hardening precipitates, can be observed. Microstructural observation gives a visual picture of such grains, while their diffraction instrumentally identifies their existence. The above mentioned analysis are complementary, together they can give a more sophisticated image about the composition.

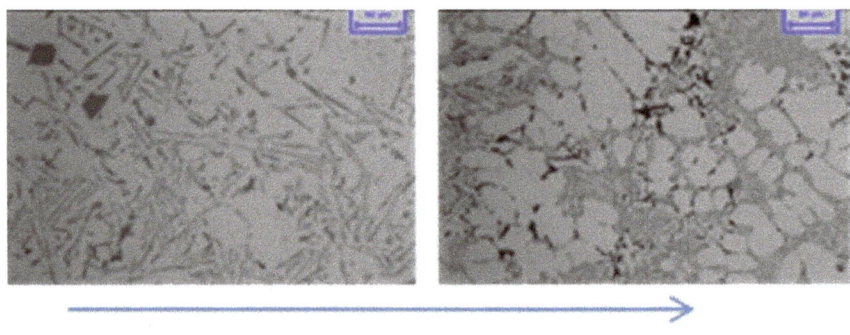

Mg content

Figure 1. Microstructure of the metallic alloy, showing the higher quantity of the intermetallic phases when the Mg content increases

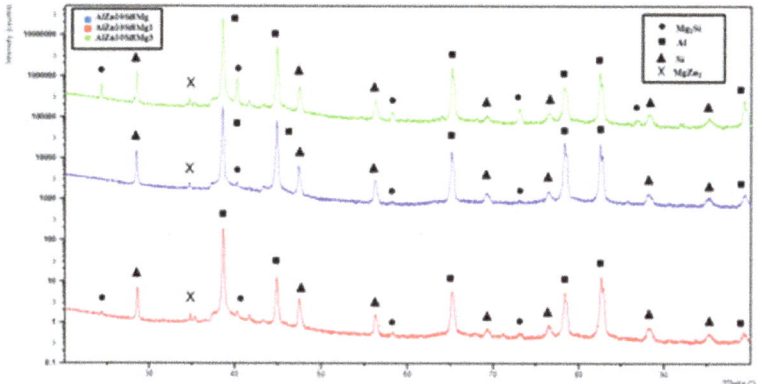

Figure 2. X-ray spectra of the investigated alloys

The attention now is moved on $ZnAl_{22}Cu_2Ti$ alloy, to check how the manufacturing route can affect the development of the grains and influence the evolution of the microstructure. Also in this case use microscope and X-ray measurement have been carried out. As from the theory, rapid cooling leads to obtain a finest microstructure, while a slow cooling produces a rough microstructure, where the grains have a larger size. The microstructures reported in Figure 3 show the morphology of the two alloys.

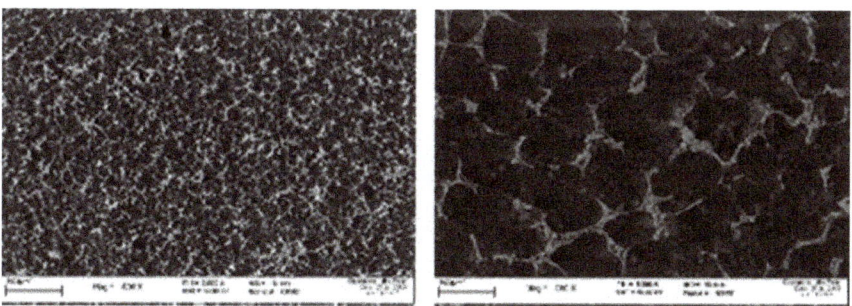

Figure 3 Microstructures of the alloy solidified in steel die (left hand side) and ceramic die (right hand side)

In X-ray spectra (Figure 4), when the peak is broadening can be considered as a sign of the lower cooling rate, and on the contrary fitted peaks are related to a rapid solidification. The manufacturing route make used of two types of dies: in one case steel made die has been employed and in the other case ceramic dies have been used. Such materials are different: one of them leading to obtain a rapid cooling (it is the case of the steel), while the other one permits to arrive to a slow solidification, because it has a better refractory property. In the last case, the alloy has more time for the development and here is a higher probability to the growth of the bigger intermetallic phases.

Other example reported here is related to a CoCr-based alloy, with a slightly different chemical composition. The structure of the alloys is different as a function of their composition and compared to a reference alloy, which is a commercially available alloy, while the other three alloys are experimentally alloys produced with an addition of a low amount of Ti, higher amount of Ti and finally with the addition of Zr.

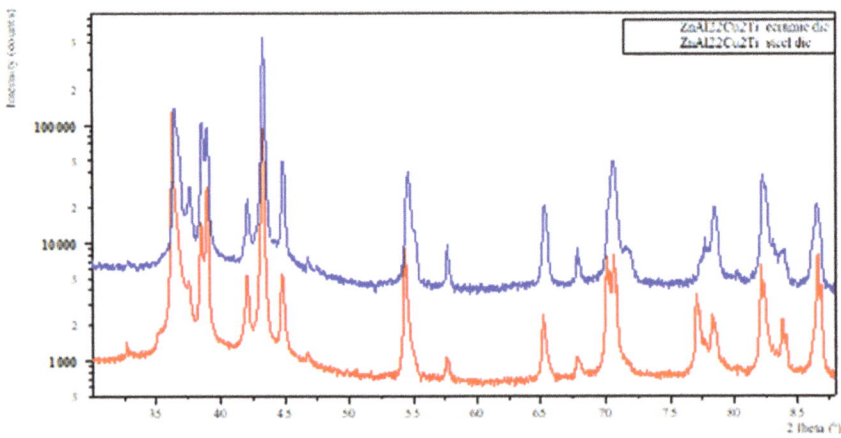

Figure 4 X-ray spectra of the alloy solidified in steel die and ceramic die

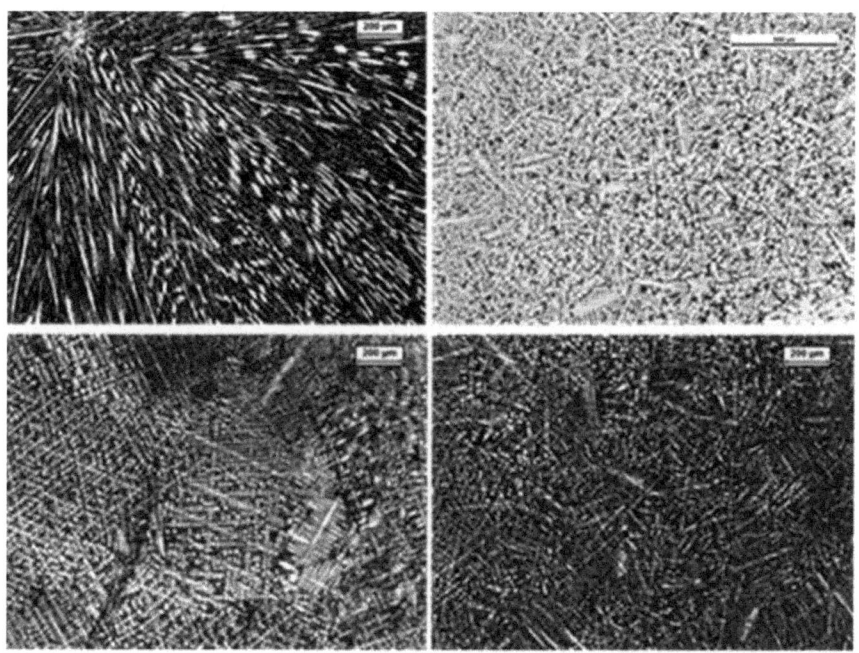

Figure 5 Microstructure of different CoCr-based alloy

As expected, their microstructural picture is different (Figure 5), they contain different grans: the alloys have different texture, since the grains are oriented in different manner. Microstructural observation and X-ray analysis (Figure 6) give the desired information about the structure of the alloy and their composition.

Let's consider the situation of duplex stainless steels and its performances. Figure 7 reports the presence of the different phases made up the microstructure. One feature which can be controlled is related to the decomposition of ferrite in σ phase and austenite. The X-Ray spectra is reported in Figure 8 and combining the two analysis performed it is possible to understand the evolution of the structure.

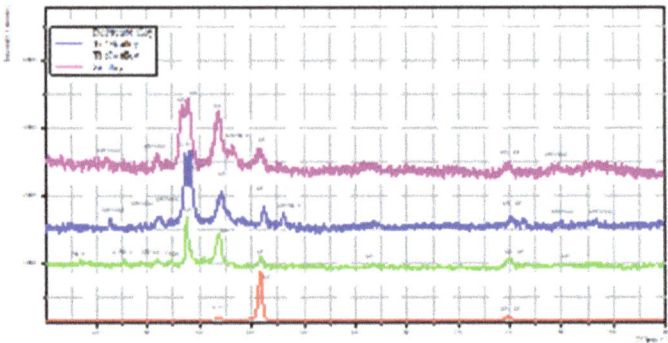

Figure 6. X-Ray spectra of the investigated CoCr-based alloy

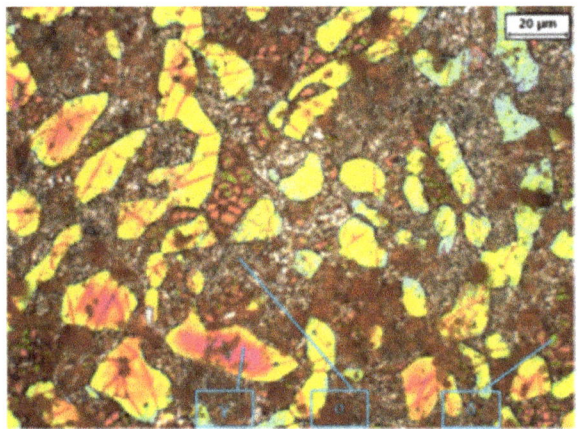

Figure 7. Microstructure of the duplex stainless steel investigated

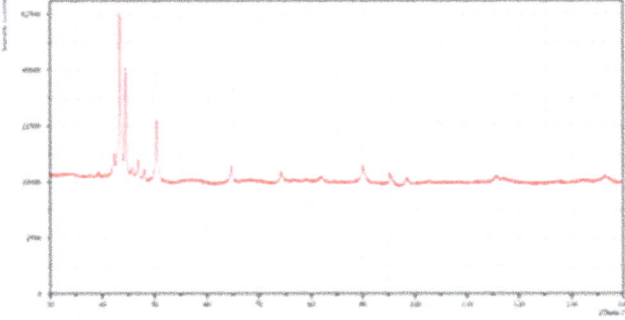

Figure 8 X-Ray spectra of the investigated duplex stainless steel

The orientation distribution function due to (i) plastic deformation, (ii) recrystallization and grain growth and (iii) phase transformations can be monitored through X-ray analysis. The material with high distribution of texture shows generally an anisotropic behavior which is correlated very well two different mechanical properties in the three possible directions. This information is important when the material has to be placed in a specific application. In case of the non-homogeneous alloy one can to take into consideration the possible solutions, as its application concerns. X-ray analysis is one of the possibility to overcome this problem and to give an instrumentally results, which together with other measurements can help to understand the real application of the metallic alloy (Figure 9).

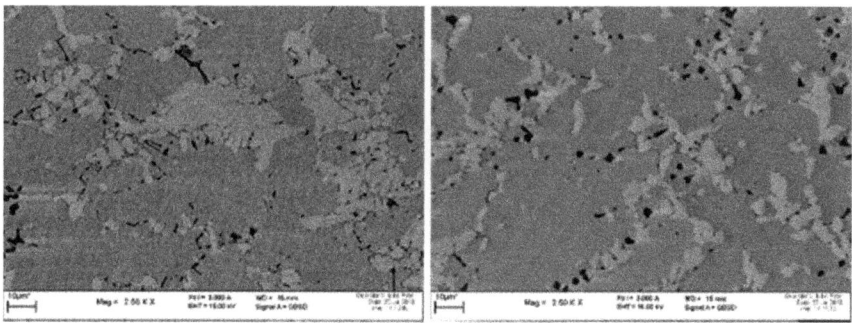

Figure 9. Microstructure of a metallic alloy showing two different topographies.

The milling time as a function of the grains size can be studied by X-ray analysis too. It is possible to appreciate, from one side, the reflections corresponding to the different crystalline structure of the alloys analyzed and from the other side it is possible to observe that the diffraction peaks broaden with increasing milling time (Figure 10), sign of a continuous variation of the grain sizes and the introduction of lattice strain within the metallic structure, which can be correlated to a higher stress content inside the alloy. Larger internal stress is directly associated to a microstructure containing higher amount of defects, minor mechanical properties, higher tendency to a micro-or macro-cracking, which involves higher probability of the failure in the same condition. After 24 h of milling, the reflection peaks (200) and (211) corresponding to α-Fe bcc disappear and the Ni peaks slightly shift towards lower angles proving that Fe atoms dissolve in the Ni lattice leading to the formation of fcc solid solution γ (Fe–Ni).

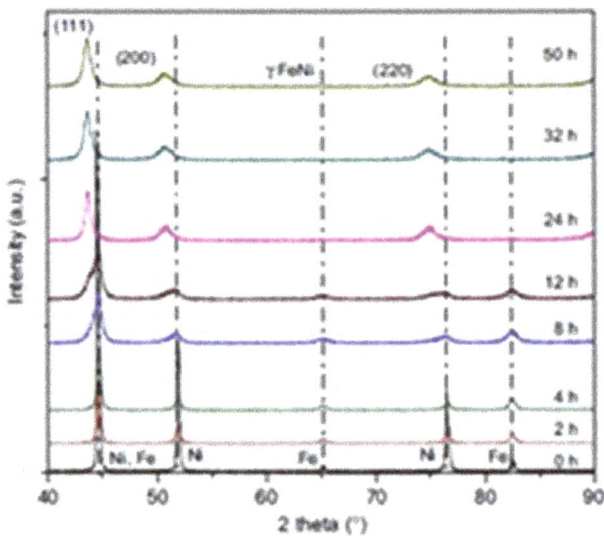

Figure 10. X-ray spectra of an Fe50Ni50 alloy with different milling time

The angular shift can be attributed to the formation of γ- (Fe–Ni) phase and also to the first-order internal stress induced by milling. The first-order angular stress acts as a macroscopic level by modifying the lattice parameter and consequently produces an angular shift in the spectra. The evolution of the average crystallite sizes and the mean level of internal strains as a function of milling time is also possible to appreciate; it is observed that the crystallite size monotonously decreases as milling time increases. This decrease in D is accompanied by an increase of the mean internal strain level.

Investigation on the crystal structures is very important for the development of new alloys, i.e., "high-entropy" alloy systems with multi-principal elements observing an enhanced hardness and plasticity as a function of the crystal structures. The variation of X-ray peak intensities of the alloy system is similar to those caused by thermal effect, but the intensities further drop beyond the thermal effect with increasing number of incorporated principal elements. An intrinsic lattice distortion effect caused by the addition of multi-principal elements with different atomic sizes is expected for the anomalous decrease in XRD intensities.

Now, an example will be considered related to metallic alloy which is in different state: in as cast condition, following heat treatment (HT or TT) and after solution hardening. In the TT samples we have no signs related to the presence of $Nd_{0.43}Y_{0.57}/Mg_5Nd$ particles and the $Mg_{41}Nd_5$ phases (Figure 11). These phases have been dissolved during the thermal heat treatment. They still show the peaks corresponding to the presence of $Mg_{24}Y_5$ and $Mg_{12}Nd$ phases, because probably the temperature fixed for the solution treatment results to be too low. Presence of peaks associated to the presence of the β' hardening precipitates (36.52° and 81.51°).

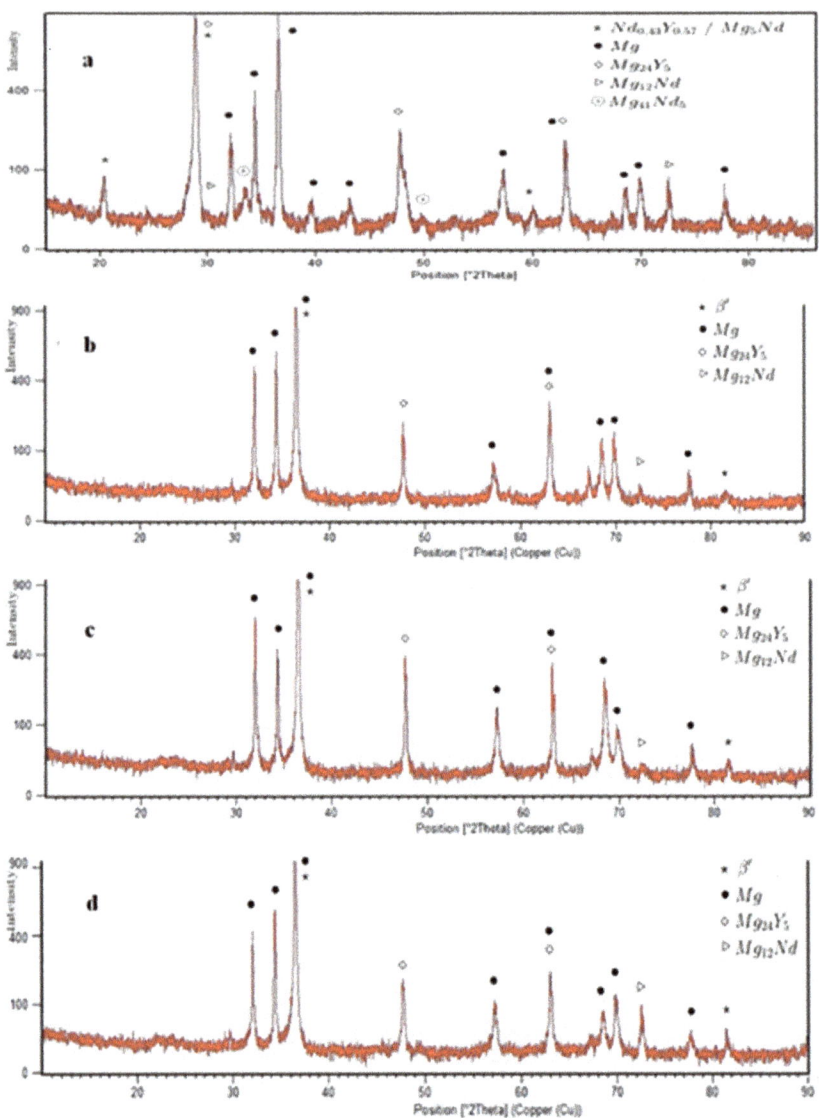

Figure 11 X-ray spectra of a metallic alloy in as cast state and after thermal treatment

7. Concluding remarks

In the present monography, the attention was focalized on giving a brief overview on different issues, such as the presence and importance of different materials which surround us, with special attention on light alloys. Challenging aspects were highlighted and some state of the art solutions were employed for illustrate how an engineer can manage the received information to obtain a better quality product.

As a final consideration about the structural analysis and more in particular on X-Ray diffraction measurement, it is correct to mention that is a very important technique which permit obtaining different information about metals and metallic alloys. The results obtained have to be compared to the results obtained by other analysis, here were compared generally to the microstructural analysis, in order to acquire additional information about the samples analyzed and in order to place it in an appropriate industrial application, where the different properties play an important role, since they determine the life-time during its application.

References

[1] Panseri C., 1966, Manuale di Fonderia d'Alluminio, III edizione, Hoepli, Milano.

[2] Nicodemi W., 2008, Acciai e leghe non ferrose, II edizione, Zanichelli, Bologna.

[3] Brisi C., 1991, Chimica applicata, II edizione, Levrotto & Bella, Torino.

[4] Campbell J., 2003, Castings, II edizione, Butterworth - Heinemann, Oxford, UK.

[5] Aguilera Luna I., Mancha Molinar H., Castro Romàn M.J., Escobedo Bocardo J.C., Herrera Trejo M., 2013, Improvement of the tensile properties of an aluminum industrial alloy by using multi stage solution heat treatments, Materials Science and Engineering A, 561, 1-6.

[6] I. Peter, M. Rosso, "Light alloys: from traditional to innovative technologies" in New Trends in Alloy Development, Characterization and Application, 3-37, InTech ed., ISBN 978-953-51-4216-4.

[7] I. Peter, B. Varga, "Some considerations on the structure refinement in Al based alloys" in Aluminium Alloys - Recent Trends in Processing, Characterization, Mechanical Behavior and Applications, InTech ed. 2017, pp. 17-38, http://dx.doi.org/10.5772/intechopen.71450.

[8] Mohamed A.M.A., Samuel F.H., Al Kahtani S., 2012, Influence of Mg and solution heat treatment on the occurrence of incipient melting in Al-Si-Cu-Mg cast alloys, Materials Science and Engineering A, 543, 22-34.

[9] Lados D.A., Apelian D., Wang L., 2010, Minimization of residual stress in heat-treated Al–Si–Mg cast alloys using uphill quenching: Mechanism and effects on static and dynamic properties, Materials Science and Engineering A, 527, 3159-3165.

[10] Sjölander E., Seifeddine S., 2011, Artificial ageing of Al–Si–Cu–Mg casting alloys, Materials Science and Engineering A, 528, 7402-7409.

[11] Ente Nazionale Italiano di Unificazione, 2010, UNI EN 1706. Aluminium and aluminium alloys - Castings - Chemical composition and mechanical properties. Milano.

[12] Manotti Lanfredi A.M., Tiripicchio A., 2001, Fondamenti di Chimica, Ambrosiana, Milano.

[13] Hultgren R., Orr R.L., Anderson P.D., Kelley K.K., 1963, Selected values of thermodynamic properties of metals and alloys, John Wiley & Sons, Inc., New York, USA.

[14] International Organization for Standardization, 1990, ISO 9227. Corrosion tests in artificial atmospheres – Salt spray tests. Geneva.

[15] Smith W.F., Hashemi J., 2008, Scienza e Tecnologia dei Materiali, III edizione, McGraw-Hill, Milano.

[16] Ente Nazionale Italiano di Unificazione, 1992, UNI EN 10045-1. Materiali metallici - Prova di resilienza su provetta Charpy - Metodo di prova. Milano.

[17] R. Douglas, D. Kuhlmann, Guidelines for precision hot forging with applications, Journal of Materials Processing Technology, 98 (2000), 182-188.

[18] T. Altan, G. Ngaile, G. Shen, M. Shirgaokar, Cold and Hot Forging Fundamentals and Applications, Copyright © 2005 ASM International®, Materials Park, Ohio, Chapter 2, pp. 7-15.

[19] T. Altan, G. Ngaile, G. Shen, M. Shirgaokar, Cold and Hot Forging Fundamentals and Applications, Copyright © 2005 ASM International®, Materials Park, Ohio, Chapter 4, pp. 25-49.

[20] E.G. Thomsen, C.T. Yang, S. Kobayashi, Mechanics of Plastic Deformation in Metal Processing, The Macmillan Company, 1965.

[21] J.R. Douglas, T. Altan, "Flow Stress Determination of Metal at Forging Rates and Temperatures", Trans. ASME, J. Eng. Ind., Feb 1975, pp 66.

[22] T. Altan, F.W. Boulger, "Flow Stress of Metals and Its Application in Metal Forming Analyses", Trans. ASME, J. Eng. Ind., Nov 1973, p 1009.

[23] T. Altan, G. Ngaile, G. Shen, S. Isbir, P. Barve, Cold and Hot Forging Fundamentals and Applications, Copyright © 2005 ASM International®, Materials Park, Ohio, Chapter 13, pp. 151-157.

[24] A.N. Bruchanow, A.W. Rebelski, "Gesenkschmieden und Warmpresse", Verlag Technik, Berlin, 1955.

[25] T. Altan, G. Ngaile, G. Shen, M. Shirgaokar, Cold and Hot Forging Fundamentals and Applications, Copyright © 2005 ASM International®, Materials Park, Ohio, Chapter 10, pp. 107-113.

[26] K. Lange, Ubersicht uber Verfahren zur Erzeugung von Arbeitsflachen von Hohlformwerkzeug (Overview of process to make working surfaces of dies and molds), Ind.-And 93 (1971) 199-200 (in German).

[27] Anilchandra R. Adamane, Lars Arnberg, Elena Fiorese, Giulio Timelli, Franco Bonollo, "Influence of injection parameters on the porosity and tensile properties of high-pressure die cast al-si alloys: a review", 2015 American Foundry Society.

[28] F. Bonollo, N. Gramegna, P.Parona, "Handbook of defects in HPDC".

[29] D. Matisková, Š. Gašpar, L. Mura, "Thermal Factors of Die Casting and Their Impact on the Service Life of Molds and the Quality of Castings", Acta Polytechnica Hungarica, Vol. 10, No. 3, 2013.

[30] Ghomashchi M.R., "High-Pressure Die Casting: Effect of Fluid Flow on the Microstructure of LM24 Die-Casting Alloy", J. Mater. Process. Technol., vol. 52, pp. 193–206 (1995).

[31] Niu X.P., Tong K.K., Hu B.H., Pinwill I., "Cavity Pressure Sensor Study of the Gate Freezing Behaviour in Aluminium High Pressure Die Casting", Int. J. Cast Met. Res., vol. 11, pp. 105–112 (1998).

[32] A. Bouayad, C. Gerometta, A. Belkebir, A. Ambari, "Kinetic interactions between solid iron and molten aluminium", Materials Science and Engineering, Volume 363, Issues 1-2, 20 December 2003, Pages 53–61.

[33] Property of Motul Tech Baraldi, "I lubrodistaccanti e la lubrificazione stampo in Pressocolata", corso di formazione.

[34] Bohler W300 Hot Work Tool Steel properties.

[35] D. C. Montgomery, G.C. Runger, "Applied Statistics and Probability for Engineers", Wiley, 6th edition, 2014.

[36] C. Rosbrook"Analysis of thermal fatigue and heat checking in die-casting dies: a finite element approach", Master Degree Thesis in Mechanical Engineering, Ohio State University, 1992.

[37] A. Long, D. Thornhill, C. Armstrong, D. Watson,"Predicting die life from die temperature for high pressure dies casting aluminium alloy", Applied Thermal Engineering 44 100-107, 2012.

[38] J. L. Graff, L. H. Kallien, "The Effect of Die Lubricant Spray on the Thermal Balance of Dies", Chem-Trend, Incorporated.

[39] I. Peter, M. Rosso, "From hot forging to thixoforging: FEM analysis of thixoforging process for steering piston production" Solid State Phenomena Volume 217-218, 2014, Pages 366-373.

[40] M. Rosso, I. Peter, "Defect control on Al castings for excellent quality and improved performances through novel Rheocasting processes", TMS Linking Science and Technology 2012, March 11-15, 2012, Orlando (USA);

[41] I. Peter, M. Rosso, "The effects of microstructural characteristics and casting defects on the mechanical failure of

Al-based alloys", Proc. 12th International Conference on Aluminium Alloys (ICAA'12), Set. 5-9, 2010, Yokohama- Japan;

[42] I. Peter, M. Rosso, C. Castella, "Investigations on coating of dies for advanced squeeze casting process" Acta Metallurgica Slovaca, vol. 20 n. 1, 2014, pp. 18-27. - ISSN 1335-1532.

[43] I. Peter, M. Rosso, F.S. Gobber, "Study of protective coatings for aluminum die casting molds", Applied Surface Science 358 (2015) 563–571.

[44] A. Guittoum *et al.* "X-ray diffraction, microstructure, Mossbauer and magnetization studies of nanostructured Fe50Ni50 alloy prepared by mechanical alloying", Journal of Magnetism and Magnetic Materials 320 (2008) 1385–1392.

ARA PUBLISHER

The most recent books published by the American Romanian Academy of Arts and Sciences

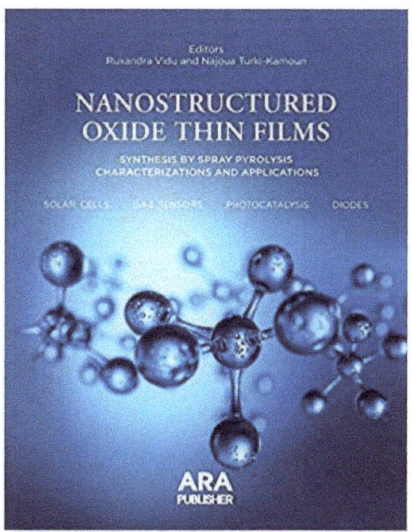

NANOSTRUCTURED OXIDE THIN FILMS SYNTHESIZED BY SPRAY PYROLYSIS.: CHARACTERIZATIONS AND APPLICATIONS
Paperback: 290 pages
Publisher: American Romanian Academy of Arts and Sciences (March 1, 2018)
Language: English
ISBN-10: 1935924249
ISBN-13: 978-1935924241
Product Dimensions: 7.4 x 0.6 x 9.7 inches

TREATISE OF MEDICAL, MORPHO-FUNCTIONAL, MOTRICITY, CULTURAL AND META-PSYCHOLOGICAL ANTHROPOLOGY
Paperback: 526 pages
Publisher: American Romanian Academy of Arts and Sciences (December 20, 2018)
Language: English
ISBN-10: 1935924273
ISBN-13: 978-1935924272
Product Dimensions: 8.3 x 1.1 x 11.7 inches

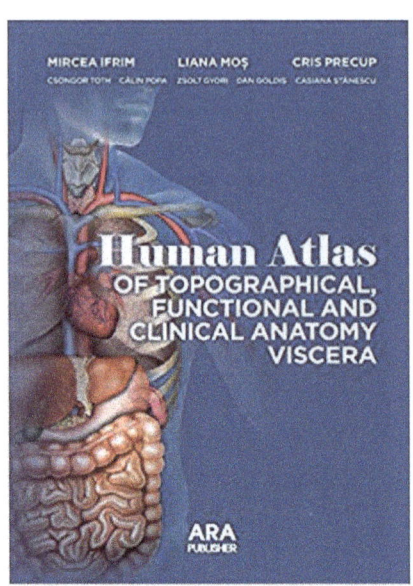

HUMAN ATLAS OF TOPOGRAPHICAL, FUNCTIONAL AND CLINICAL ANATOMY VISCERA
Paperback: 336 pages
Publisher: American Romanian Academy of Arts and Sciences (June 30, 2016)
Language: English
ISBN-10: 1935924206
ISBN-13: 978-1935924203
Product Dimensions: 8.3 x 0.7 x 11.7 inches

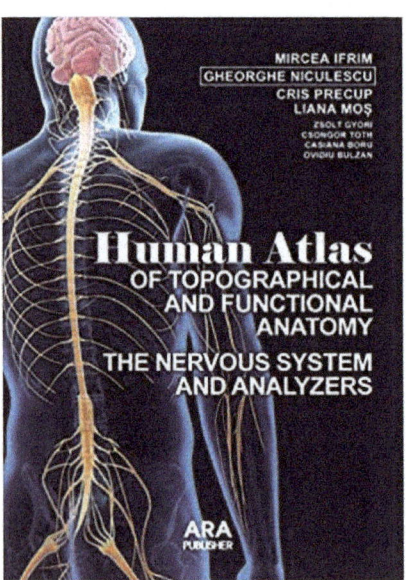

HUMAN ATLAS OF TOPOGRAPHICAL AND FUNCTIONAL ANATOMY: THE NERVOUS SYSTEM AND ANALYZERS
Paperback: 382 pages
Publisher: American Romanian Academy of Arts and Sciences (October 27, 2017)
Language: English
ISBN-10: 1935924222
ISBN-13: 978-1935924227
Product Dimensions: 8.3 x 0.8 x 11.7 inches

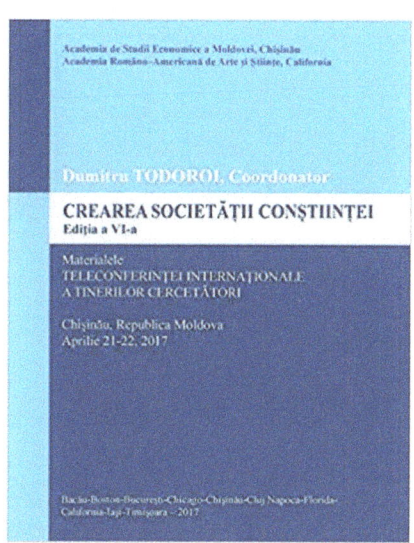

HUMAN ATLAS OF TOPOGRAPHICAL AND FUNCTIONAL ANATOMY: THE NERVOUS SYSTEM AND ANALYZERS
Paperback: 382 pages
Publisher: American Romanian Academy of Arts and Sciences (October 27, 2017)
Language: English
ISBN-10: 1935924222
ISBN-13: 978-1935924227
Product Dimensions: 8.3 x 0.8 x 11.7 inches

CLASH OVER ROMANIA: BRITISH AND AMERICAN POLICIES TOWARD ROMANIA: 1938-1947 (VOL. II)
Paperback: 303 pages
Publisher: American Romanian Academy of Arts and Sciences (July 11, 2014)
Language: English
ISBN-10: 9781935924166
ISBN-13: 978-1935924166
Product Dimensions: 6 x 0.6 x 9 inches

ARA Publisher, an international Academic Press of the
American Romanian Academy of Arts and Sciences
University of California Davis
http://www.AmericanRomanianAcademy.org
Email: *info@AmericanRomanianAcademy.org*
Address: P.O. Box 2761, Citrus Heights, CA 95611-2761

www.ingramcontent.com/pod-product-compliance
Lightning Source LLC
Chambersburg PA
CBHW050929240426
43671CB00019B/2965